Division of Policy Analysis Services

1 Economic Development in the States

State Business Incentives and Economic Growth: Are They Effective? A Review of the Literature

by Roger Wilson, Policy Analyst
The Council of State Governments

© **Copyright, The Council of State Governments, 1989**
ISBN 0-87292-090-9
C-140
$20.00

State Business Incentives and Economic Growth

Contents

Foreword iv
Acknowledgments iv
Preface v

Historical Overview 2
The Nature and Process of Plant Location Decisions 8
Empirical Studies 12
  Survey Research 12
  Statistical Approaches 17
The Increase in Incentives — Why Has It Happened? 22
Concluding Observations 27

Notes 29
Bibliography 35

Tables:
1. Growth in State Financial Incentive Programs 3
2. State Tax Incentives for Business:
   Changes Between 1977-88, 1984-88 5
3. State Financial Incentives for Business:
   Changes Between 1977-88, 1984-88 6
4. Incentives Offered to Attract Major Automobile Plants .. 8
5. Factors Mentioned in "Summaries"
   Ranked by Index of Importance 13
6. Comparative Importance of Factors in Locating Plants .. 15

Foreword

The past few years have witnessed a tremendous increase in state programs designed to stimulate economic growth. We should not be surprised by this trend.

Felix Frankfurter said that a nation's economy forms the very foundation of a people's social and moral well-being. State officials certainly are aware of the importance their constituents place on this "foundation." Many have discovered that voters cast their ballots as much in review of the economy as in judgment of the officeholder.

This report, by Roger Wilson of the Division of Policy Analysis Services, reviews the literature on the effectiveness of one of the economic development tools — business incentives — on economic growth in the states.

It is the first in a series of three volumes produced by Policy Analysis Services as a result of its study of state business tax and financial incentives. The second report in this series, *The Changing Arena: State Strategic Economic Development*, focuses on the states' response to changes in their economic environment. The third, *The States and Business Incentives: An Inventory of Tax and Financial Incentive Programs*, provides a state-by-state description of formal programs designed to create, expand and recruit business and industry.

Taken as a group, we believe these reports will provide an invaluable resource for state policy-makers considering various economic development options.

May 1989

Carl W. Stenberg
Executive Director
The Council of State Governments

Author's Acknowledgments

Like most reports, without the assistance of many people this one would not have been possible. The report owes much to our research team — Susan Stone and Jan Norris Clarke — who spent countless hours in the University of Kentucky's library searching out numerous articles and documents, some obscure, for the bibliography. I wish to thank Lee Walker who helped formulate many of the issues that needed to be researched. Keon Chi, Elaine Knapp and Norm Sims read through the entire report and made invaluable suggestions, as well as criticisms. Nancy Olson not only had the laborious task of transforming "scribbled" sheets into legible pages of type, but also suggested appropriate editorial corrections. Finally, I am deeply indebted to Debbie Gona, director of the Division of Policy Analysis Services, who had the unenviable and tedious task of simultaneously maintaining editorial consistency and providing counsel (and sometimes good humor). She performed this job with such perfection that I am tempted to blame her for any deficiencies in the final product. Only gratitude dissuades me.

I thank all of these individuals for their invaluable contributions and wish to make clear that any interpretations or factual errors in this report are solely mine.

Preface

Business tax and other financial incentives used by states and communities to stimulate economic growth and create job opportunities are among the more controversial tools of economic development, particularly as a result of some of the headline-making incentive packages offered to auto plants and other large firms in recent years.

Proponents argue the use of incentives helps create a favorable business climate and that an attractive package of incentives is often the ammunition a state needs to compete in bidding wars against other states for industry. Opponents criticize the overuse of tax concessions and other fiscal inducements for their ineffectiveness and inefficiency, arguing that their costs far outweigh any long-term benefits.

State policy-makers face a fundamental challenge as they decide whether and how to tailor incentives to attract specific firms or to offer incentives as part of more comprehensive economic development strategies. To help them make more informed decisions, The Council of State Governments' (CSG) Policy Analysis Services undertook a study in 1988 related to state business tax and financial incentives.

Preliminary explorations of the issue, coupled with escalating media attention, suggested states *had* increased their use of incentives to attract business and industry. But what were the states offering in the way of formal tax and financial incentive programs? And were their numbers really increasing or were the headlines exaggerating the issue by grabbing on to some of the larger packages offered to firms?

More to the point, though, was there solid evidence that business incentives, as policy instruments, were effective in spurring economic growth in the states? Was the presence of business incentives even a factor in business investment and location decisions? If not, what forces were driving the expansion of their use? And what effect would the changing economic environment – one in which states are increasingly forced to compete in an international arena – have on the use of incentives and on economic development strategies in general?

The results of this undertaking, while revealing, do not offer any easy, uncomplicated answers for the states. However, they do offer a new look at some of the old assumptions surrounding the use of business incentives as a method for stimulating economic growth, as well as insight into some of the more recent economic development approaches and activities across the states.

Over the years, the number of states offering business incentives through formal programs *has* steadily increased. In fact, during the past five years a majority of the states offered a variety of tax and financial incentives for business and industry. Most states reduced their overall tax rates or tax rates for specific businesses. Some offered "customized" tax incentive packages for selected firms or projects. And to further assist businesses, most states initiated various financial incentive programs, apart from those already offered through existing federal programs.

And yet, a comprehensive review of past studies on the effects of incentives reveals *no* statistical evidence that business incentives actually create jobs. What those studies do suggest, overall, are some contradictory findings on the significance of incentives: they *are not* the primary or sole influence on business location decision-making and, relative to other factors, they *do not* have a primary effect on state employment growth; but they *do* become more effective when all other variables are equal among competing sites within a region or substate area, and they *are* important in that they often are used as a component in business climate indices.

Even though incentives cannot be linked to effecting tangible improvement in state economic growth, they apparently remain important, "psychological," or political weapons. Some analysts argue that states have a common fear of being "outbid," and that states, in what amounts to an "arms race" mode of behavior, are forced into matching and beating each other's offerings, using incentives as defensive measures against their competitors.

But the field of "competitors" has expanded in the new economic environment. No longer is the competition just a neighboring state or region; now, it is national and international in scope, and states are responding to that new reality. For example, a 1988 CSG survey of state economic development agencies revealed that the packaging of business tax and financial incentives remains an important issue and that states believe those inducements can have a significant effect on new business investment and job creation. Increasingly, however, states are using incentives within the context of a comprehensive economic development strategy that accounts for their strengths and weaknesses with regard to regional, national and international competition.

Indeed, the new global economic environment that has emerged over the last decade is limiting the states' capacity to pursue economic development in older, more conventional ways. The changes that created the economic restructuring of the 1980s are likely to accelerate, and states will need to reconsider their approaches to economic development and adjust their policies to survive and prosper in the world economy. That is the reality for state policy-makers as the decade of the 90s draws nearer.

This is just one volume in a series of three companion pieces produced by Policy Analysis Services as a result of its 1988 study: *State Business Incentives and Economic Growth: Are They Effective? A Review of the Literature*; *The Changing Arena: State Strategic Economic Development*; and *The States and Business Incentives: An Inventory of Tax and Financial Incentive Programs*.

Each provides insight into one piece of this economic development puzzle. Together, they offer a set of perspectives on where the future is likely to lead the states in their economic development efforts and in their interactions with each other and the rest of the world.

Deborah A. Gona
Director, CSG Policy Analysis Services

State Business Incentives and Economic Growth: Are They Effective?

A Review of the Literature

To most state policy-makers, economic growth[1] represents the best barometer of an effective economic development policy. A healthy economy usually results from policies that generate business activity in the state — activity that creates employment opportunities and expands the state's revenue base. State policy-makers are interested in creating and implementing economic development policies that attract industry and encourage retention and expansion of existing businesses. Their concern, however, centers on the types of economic development policy instruments needed to accomplish these objectives.

One of the most controversial types of policy instruments employed by states is the array of tax and fiscal inducements designed to attract new businesses or retain existing ones. Proponents argue that business incentives[2] — such as tax concessions, loans, gifts and industrial development bonds — reduce business costs; aid capital expansion, particularly for existing firms; improve efficiency in resources allocation (especially in areas where structural unemployment and uniform wage rates exist); symbolize a favorable attitude toward business; and yield profitable rates of return on public investments.[3] Opponents argue that business incentives are inefficient, inequitable and costly; do not influence business investment or location decisions; and merely foster "bidding wars" between states.[4] Although the literature on the effectiveness of business incentives is mixed, the use of these economic development tools continues to escalate.[5]

Thus, an understanding of the effectiveness of business incentives should be important to state officials who want to develop policy instruments that will spur economic growth. But it should be equally important to determine the factors that do influence business investment decisions: What factors and processes do businesses consider and use to make investment and locational decisions? Have technological and market forces changed the types and modes of business investment? Do high taxes stifle growth both in business activity and employment opportunities? Does the use of business incentives infer a favorable business climate? What forces drive the expansion of business incentives?

This review of the literature on the effects of business incentives on state economic growth is designed to assist state policy-makers in their decisions regarding the development of these tools. The first section presents a brief overview of the use of state business incentives. The second examines the nature of and processes underlying business location decisions. The third section discusses the findings

Historical Overview

State financial assistance to businesses is not a recent phenomenon.[6] In 1791, New Jersey, in an effort to stimulate private investment in the state, granted a tax exemption to a new manufacturing company owned by Alexander Hamilton.[7] As early as 1800, states were providing capital to private industries and financing infrastructure to promote economic development. According to DeWitt John, "By 1844, Pennsylvania, for example, had invested over $100 million and had placed directors on the boards of over 150 mixed corporations [and] . . . state and local governments provided over 55 percent of the initial capitalization for the Southern Railroad."[8]

During the 1930s, Southern states, suffering from the effects of the boll weevil invasion, the Depression and New Deal farm legislation, devised a new and aggressive industrial recruitment strategy that expanded financial assistance to businesses.[9] This assistance centered around tax incentives, industrial development bonds, loans, "gifts" and advertising.[10] The aim was to attract large companies from other states by reducing the capital cost of plant location.[11] In 1936, Mississippi pioneered the use of business incentives to attract industry and, at the same time, became the first state in the nation to authorize local industrial bond financing.[12] Under its "Balance Agriculture with Industry" program,[13] Mississippi used tax-exempt bonds to reduce the capital cost associated with the construction of new plants.[14] In assessing the South's use of business incentives during this period, James Cobb noted:

> These incentives were designed to give each state an edge in dealing with prospective industries, but the use of these devices quickly became so common throughout the region as to negate this advantage to any individual state . . . As competition for industry became more heated, public officials from governors down to the city councilmen were expected to demonstrate a wholehearted commitment to economic development.[15]

Caution and World War II slowed the development of business incentives, but their use blossomed again in the 1950s and 1960s.[16] Benjamin Bridges,[17] in an article on state and local inducements for industry, tracked the rapid development of business incentives during this period. He noted that in 1949, Maine authorized the first statewide business development corporation. Ten years later, 21 states had business development corporations, and by 1963, 31 had such entities. In 1955, New Hampshire created the first state industrial finance authority to guarantee industrial loans made to industrial borrowers or make direct loans of state funds to businesses. Just

and limits of empirical studies that gauge the effects of business incentives on state economic growth. The fourth reviews the research findings on the forces motivating the use of business incentives. The final section offers observations about these findings and their implications for state economic development policy. An updated bibliography on the business incentive literature is included at the end of this report.

eight years later, 19 states had these authorities. From 1946 to 1963, 17 states authorized tax concession programs. Table 1 shows the growth in various types of financial incentive programs from the late 1950s to the early 1960s.

The real explosion of business incentives, however, came as an aftermath of the employment crisis of the 1970s and the recession of the early 1980s. As states tried to relieve unemployment and bolster revenues that eroded during the recession, their use of business incentives to attract industrial prospects accelerated. States became so competitive in offering business incentives as part of their recruitment battle for new or relocating plant facilities that some commentators referred to the flurry of activity as "the Second War Between the States."[18] Tables 2 and 3 show the increase in the use of state business tax incentives and other financial incentives, respectively, from 1977 through 1988 and from 1984 through 1988.

By the early 1980s, every state offered some form of tax and financial incentives. In 1981, states spent an estimated $20 billion on financial assistance to businesses, which included incentives designed to attract new businesses or retain existing ones.[19] Table 4 shows the dollar amounts some states offered in the 1980s to attract and win major automobile plants. These amounts, however, do not include the cost to states that were unsuccessful in their bids to attract these industries. As Barry Rubin and Kurt Zorn pointed out:

> The process of creating, packaging, and promoting these incentives is by no means costless. As such, the short run costs become even greater when one adds the cost of those incentives which failed to attract the sought-after industries.[20]

Many scholars and practitioners believe that this "War Between the States" has abated.[21] According to this view, the state's use of business incentives no longer focuses merely on industrial recruitment, but also on retention and expansion. They argue that states are concentrating more on strategic planning, infrastructure improvements, education and global competitive market strategies. Thus, R. Scott Fosler wrote:

Table 1
Growth in State Financial Incentive Programs

	Number of states with program	
	1959	1963
State Industrial Finance Authorities	5	19
State Loan Guarantee Programs	4	7
State Direct Loan Programs	3	14
Business Development Corporation	21	31
Local Industrial Bond Financing	13	25

Source: B. Bridges, "State and Local Inducements for Industry: Part I," *National Tax Journal*, 18 (March 1965): pp. 1-14.

States have become leaders in confronting the global challenges to American competitiveness.... Until the 1970s, the few states that had formal economic development programs focused them on efforts to attract business to the state. In response to the economic turbulence of the 1970s and especially since the severe recessions of the early 1980s, many states have broadened their efforts to include creation, expansion and retention, as well as attraction of business. They have rediscovered the economic importance of such traditional services as education and transportation. And they have established new programs in numerous areas: for example, to provide new sources of capital, to promote new technologies, to support small business and entrepreneurship, and to expand export markets. This surge of activity has intensified throughout the 1980s and shows no signs of abating.... In the past, American regions have traded and competed almost exclusively with one another.... Today, by contrast, the regions of America are trading and competing not only with one another but with the regions of the world.[22]

Despite the changing perspectives on state economic development strategies in general and business incentives in particular, the number of states using business incentives has increased and the types and number of business incentives offered by states have expanded.[23] Moreover, "smokestack chasing" — the use of business incentives to lure domestic and international manufacturing companies, especially automobile plants and their suppliers — is being replaced by what one state official has called "third wave investment"[24] or the recruitment of foreign high-tech oriented industries:

With Japan's U.S. investment binge showing no signs of letting up, the competition is growing so heated that almost no concession is out of bounds. For years only a handful of states — Tennessee, Kentucky, Michigan, Illinois and Georgia — consistently bid for new Japanese manufacturing plants. Now more and more states see the benefits of courting long-term foreign employers that pay above average wages. States in the Pacific Northwest and the Deep South are joining the plant-site auction; in all more than 30 states have opened offices in Tokyo to recruit businesses.[25]

Arguably, state economic development strategies are focusing on retention and expansion of existing businesses, and states are developing approaches to compete in a global economy. However, as Dennis Grady notes, "because of the established precedent of direct business subsidization, companies can essentially blackmail states with the threat of leaving or generating competitive bidding between states over a new investment."[26] A number of states have acquiesced to the use of tax and financial incentives in order to retain companies threatening to leave. Recently, for example, the Nebraska Legislature overhauled the state's tax structure after Con Agra, Inc., threatened to move its corporate headquarters.[27]

Whether their goal is to attract or retain businesses, states are spending huge sums of monies from limited economic development coffers. These expenditures place small existing businesses at a dis-

Table 2
State Tax Incentives for Business:
Changes Between 1977-88, 1984-88

Types of Tax Incentives	1977	1984	1985	1986	1987	1988	Changes Between 1977-88	Changes Between 1984-88
Corporate Income Tax Exemption	21	28	31	33	33	31	+10	+3
Personal Income Tax Exemption	19	22	24	26	27	28	+9	+6
Excise Tax Exemption	10	16	16	18	19	19	+9	+3
Tax Exemption or Moratorium on Land, Capital Improvements	23	32	34	33	34	35	+12	+3
Tax Exemption or Moratorium on Equipment, Machinery	28	32	34	35	35	39	+11	+7
Sales/Use Tax Exemption on New Equipment	33	38	42	42	44	44	+11	+6
Tax Incentive for Creation of Jobs	n/a	27	30	31	32	33	n/a	+6
Tax Incentive for Industrial Investment	n/a	24	29	29	30	32	n/a	+8
Tax Exemption to Encourage Research and Development	9	19	22	24	25	25	+16	+6

n/a = not available

Source: Compiled from January/February 1978, January/February 1985, August 1985, October 1986, October 1987 and October 1988 issues of *Site Selection and Industrial Development*, Conway Data, Inc.

Adapted from Keon Chi, *The States and Business Incentives: An Inventory of Tax and Financial Incentive Programs* (Lexington, Ky.: The Council of State Governments, 1989).

advantage since they provide the large recruited companies with excess profits, gained from business subsidies, which could not be obtained without some risk in the marketplace. And giving tax and financial advantages to new firms might antagonize those existing, competing firms. In 1985, Washington state landed a high-tech firm after granting it tax concessions that did not apply to existing firms. One of the existing firms, Immunex Corp., a biotechnology company, vowed that it would leave the state when the time came to build a new plant "unless there is a tax overhaul."[28] Business incentives might reduce the overall amount of monies

Table 3
State Financial Incentives for Business:
Changes Between 1977-88, 1984-88

Types of Financial Incentives	Year						Change Between 1977-88	Change Between 1984-88
	1977	1984	1985	1986	1987	1988		
	Number of States							
State Authority or Agency Revenue Bond Financing	20	37	41	42	44	44	+24	+7
State Loans for Building Construction	19	30	34	35	36	38	+19	+8
State Loans for Equipment, Machinery	13	27	33	34	35	37	+24	+10
State Loan Guarantees for Building Construction	14	20	26	22	22	25	+11	+5
State Loan Guarantees for Equipment, Machinery	13	21	21	23	23	26	+13	+5
State Financing Aid for Existing Plant Expansion	29	37	39	38	41	42	+13	+5
State Incentive for Establishing Industrial Plants in Areas of High Unemployment	17	24	25	27	31	31	+14	+7

Source: Compiled from January/February 1978, January/February 1985, August 1985, October 1986, October 1987 and October 1988 issues of Site Selection and Industrial Development, Conway Data, Inc.

Adapted from Keon Chi, *The States and Business Incentives: An Inventory of Tax and Financial Incentive Programs* (Lexington, Ky.: The Council of State Governments, 1989).

available for other economic development projects, entrepreneurial ventures and economic development tools and services (e.g., education, worker training and infrastructure improvements). And given the uncertainties of an increasingly competitive world economy, there is no guarantee that a company which receives millions of dollars in state financial assistance will remain in the state, as exemplified by the recent Volkswagen plant closing in Pennsylvania.[29] Ironically, given the new focus on retention and expansion of existing business and the emphasis on global competitive strategies, states still seem unable to break away from the use of business incentives. As William Fulton said:

Despite the sense of importance such comments convey, governors cannot be expected to abandon the smokestack hunt, even when they recognize that their main economic development efforts should lie elsewhere. They will still see the possibility of a political slam dunk in luring a big industrial plant into an area where local residents are desperate for high-paying, blue-collar jobs... it would seem more likely that states chasing auto plants would understand the subtle economic trends in the auto industry: a worldwide glut of the auto production capacity and the likely consolidation of the industry into a few worldwide giants such as GM, Ford and Toyota... But these trends didn't stop the governors of seven states from appearing together on Phil Donahue's television show in hopes of luring GM's Saturn operation, which is in essence a start-up company. They didn't stop Illinois... from using a subsidy of about $30,000 per job to lure Diamond-Star Motors, a joint venture between two smaller auto manufacturers, Chrysler and Mitsubishi, that industry analysts say are not as well equipped to survive an expected worldwide shakeout. Nor did it prevent Indiana from coughing up $50,000 per job — about the same as Toyota's Kentucky package — to bring a joint Fuji-Isuzu plant to Lafayette.[30]

To summarize:
- There are drastically more states using business incentives. Currently, every state offers some form of incentive, ranging from tax concessions to outright cash payoffs.
- Some analysts argue that states now use incentives for retention and expansion, in addition to industrial recruitment. States are focusing more attention on strategic planning, infrastructure, education and global competitive market strategies.
- Yet indications are that even with a new emphasis on developing global competitive strategies, states are unable or unwilling to wean themselves from the use of business incentives.

A state's decision to offer incentives to prospective companies stems from the notion that these incentives will influence a company's location or expansion decision. Various studies outlining industrial location theories might help explain some of the elements that influence where a company will locate or expand. It is important to recognize, however, that such abstractions often ignore "real world" considerations that could drastically alter the "ideal" location decision. Even so, familiarity with these theories might prove valuable to state policy-makers in identifying the major location factors of industry and providing guidelines for the development of business incentive programs.

The Nature and Process of Plant Location Decisions

Historically, transportation cost was considered the most important single determinant of plant location. An early articulator of this approach, Alfred Weber,[31] speculated that the best place for a plant to locate was wherever the cost of transporting raw materials to the plant and transporting products to its market was the least. Deviations would occur only when savings in labor costs and "agglomeration economies" (such as access to markets or proximity to auxiliary firms) exceeded additional transportation costs; subsequently, a new "best place" would be identified. However, Weber's theory assumed a market where perfect competition between buyers and sellers existed. Other advocates of this "least-cost" theory of manufacturing made some improvements and modifications in Weber's model, but still emphasized the essential role of transportation costs in plant location decisions.[32]

Some of the shortcomings of the "least-cost" theory of plant location were addressed by other analysts. While Weber assumed that

Table 4
Incentives Offered to Attract Major Automobile Plants

Company and Location	State Investment	Estimated or Actual Start Up
Nissan Motor Co Smyrna, Tenn.	$ 33 million	1983; expansion 1985
Mazda/Ford Motor Flat Rock, Mich.	$ 48.5 million[1]	September 1987
Toyota Motor Corp. Scott County, Ky.	$150 million[2]	1988
Saturn Corp. Springhill, Tenn.	$ 80 million	Phase I by 1990
Diamond-Star Motors Bloomington/Normal, Ill.	$ 83.3 million	Spring 1988
Fuji-Isuzu Lafayette, Ind.	$ 86 million[3]	1989

Source: H. Brinton Milward and Heidi Hosbach Newman, "State Incentive Packages and the Industrial Location Decision," paper presented at the Southern Political Science Association Meeting, Charlotte, NC, November 5-8, 1987.
[1] Michigan also posted a $32 million pollution control bond.
[2] Miscellaneous fees are estimated at an additional $9 million (i.e., legal fees, an impact survey) and bond interest payments are estimated at between $166.7 and $224 million.
[3] Estimates of Phase II total $25 million in additional state support.

the location decision rests solely on cost, Harold Hotelling[33] recognized that there is a sales advantage at alternative sites. He argued that since buyers purchase products from the company whose prices are the lowest, the best location for a company is wherever it can influence the largest number of buyers with the price that yields the greatest return. Under Hotelling's model, the size of a company's market area is mutually dependent upon competing companies located within the area.

Melvin L. Greenhut[34] restated this notion of "locational interdependence" among companies. He argued that Weber failed to recognize that a company's location decision involves more than merely substituting cost factors at alternative locations. It also involves demand factors at different locations. Greenhut suggested that this variability in market demand not only influences location decisions, but also significantly affects, and is affected by, the interdependence of firms.

A second theory of plant location was developed by August Losch[35] who argued that companies choose locations that maximize their profits, something they accomplish by extending their market areas. He recognized that a location with the lowest transportation, wage or sales costs is not necessarily a company's best site. Instead, he said a company's location decision is based on production cost factors at alternative locations and market areas a company controls from each location site.[36]

However, some analysts have argued that these theories ignore the realities of the marketplace: factors like the various actors and institutional costs involved in a company's location decision[37] and even the behavior of buyers and sellers.[38] And advancements in industrial technology have lessened the explanatory power of these theories. Transportation cost is still a major factor in the location of many industries, but technology has diminished its impact.[39] The refinement of raw materials within companies, the substitution or interchange of material inputs, efficient utilization of materials and the changing composition of industries not only has reduced transportation costs, but also created more mobile, or "footloose," firms.[40]

Moreover, many analysts are beginning to realize that companies consider other factors besides profits in determining plant locations. Non-economic factors, such as infrastructure, education, climate and better services, are becoming increasingly important in this age of high technology. And for new businesses, other, less tangible factors might take precedence over profit motive in some location decisions. For example, according to John Blair and Robert Premus:

> *New businesses are less sensitive to the profit maximizing aspects of locational choices than branch plants. Businesses in the formative stage appear to locate in the area where the founder lived implying that personal factors take precedence over strict profit maximization.*[41]

These analysts also argue that a company's location decision is not based on one particular factor, but is actually part of a "larger corporate planning process."[42] The role of a new plant is only one component of an overall corporate strategy aimed at meeting the company's future capacity requirements or its expected shortfalls. The corporate site selection team, determined by the organizational structure of the company, considers the locational characteristics the company deems most important. According to Blair and Premus, plant location decisions involve a two-stage process:

The first stage is the choice of state or region. Over half of all locational studies make their "first cut" at a multistate level. Once the state or region has been selected, a more micro-focus will culminate in the selection of a specific community and site. The important locational factors differ between the first stage where firms are seeking a general region in which to locate and the second, more geographically focused stage. In selecting a broad region, the site selection team will focus on labor, state tax variables, climate, proximity to market and other features that may have significant interregional variation, but are similar almost everywhere within the region. Locational factors that vary at the micro-geographic level of detail such as land costs, access to major roads and schools are generally available somewhere within all major regions. These micro-factors become more important when selecting a specific site or specific community within a region. Unfortunately, many location studies do not distinguish between stages of the location process. A city may have many of the important micro-geographic features such as good education and inexpensive land that would be desirable, but because of its regional setting, the city may not be a feasible site for a particular firm.[43]

These changing perspectives about the location decision should cause state policy-makers to reexamine the importance of business incentives as components in this process. Consider, for example, some of the major factors associated with the location decisions of large manufacturing companies, such as automobile plants — transportation costs, access to markets, availability of new materials and labor costs. With the advent of new plant technology and international competition, these plants are likely to consolidate (or merge their operations), meaning that they will close plant facilities in some regions and relocate to others in order to maximize the above mentioned factors. A state's use of incentives may sweeten the pie, but will not always influence a company's decision to come, stay or leave. This point can be illustrated in some recent examples.

In the 1990s, General Electric will close down its factory in Cicero, Illinois, and spend $160 million to expand its plant in Decatur, Alabama.[44] According to General Electric's chief executives, neither productivity, wages nor business incentives were important in the plant's decision to leave Cicero. Rather, the decision was based on the need to consolidate diminishing resources and to handle new manufacturing technology. Despite these reasons, both Decatur and

Cicero developed elaborate business incentive packages. The following news account illustrates:

> 'We offered to put together a package of incentives,' said Lynn Morford, a spokesman for the Illinois Department of Commerce and Community Affairs. 'But the company said its decision wasn't based on the business climate or production'... 'Illinois,' Dunham (General Electric vice-president for manufacturing) said, 'was not in a position to really affect the decision.' He said GE didn't encourage competition between the states 'because we didn't think the state impact or local impact would make this decision change.' Once the decision was made, however, GE negotiated an extensive assistance package with state and local governments in Alabama... The state [Alabama] has made a $10 million construction loan at 2 percent interest and the city will add $20 million in financing through industrial revenue bonds... In addition, Alabama will spend more than $700,000 to train about 900 new employees and an undetermined amount to rebuild roads near the plant.[45]

Thus, Illinois' attempt to retain the General Electric plant failed although it offered to develop a package of financial incentives, and Alabama developed a package of financial incentives even though General Electric would have located in that state in spite of its incentive package. A similar event occurred with the recent closing of a Volkswagen plant in Pennsylvania. In this case, however, Pennsylvania had spent more than $70 million to lure the plant from Ohio. Global competition forced Volkswagen to consolidate its operations and close the Pennsylvania plant.

> 'The location is excellent; the cost of labor is high, admittedly,' says Bahn, the [Volkswagen] Westmoreland spokesman, but Bahn added, 'if VW had asked the employees to work for nothing, it would have made no difference. It finally became obvious that our market is not going to expand for the size car we make. It's nobody's fault. Nobody here made a mistake.'[46]

Another implication of the changing theoretical perspectives on location decision-making is the changing structure of the economy. This information-age economy has witnessed the proliferation of corporate headquarters and high technology firms with a striking feature — their mobility. Although many of these companies are making location decisions on the basis of a state's universities, work force literacy levels and infrastructure developments, some are literally forcing states into awarding business incentives,[47] while a number of these "footloose" companies are taking state financial incentives and then departing.[48]

As Cobb observed:

There is a grim irony in the fact that the South, having worked so diligently to create a business climate attractive to footloose industries should now find its economic future threatened by an increase in industrial mobility. Industries fleeing the South are purchasing one-way tickets to Taiwan and other exotic destinations just as readily as they used to depart Akron, Ohio, for Opelika, Alabama. . . . The high-tech plant manager who threatens to flee Massachusetts if his demands were not met was not just whistling Dixie when he warned that he could load the entire plant on trucks and have it in North Carolina the next morning. The problem for the South is that he could just as easily follow the recent example of a General Instrument Incorporated plant that left Chicopee, Massachusetts, and without even stopping in North Carolina to refuel, headed straight to Mexico.[49]

Even so, if the purpose of a state's economic development policies is to reduce the cost factor associated with a company's location decisions, then policy instruments such as tax concessions, fiscal subsidies, loans and development bonds should increase the manufacturing activities within that state, particularly the location of footloose firms. If these incentives reduce the cost of borrowing, raise profits and increase the availability of capital, then the incentives should become an important factor in a company's decision to locate or relocate and thus, stimulate a state's economic growth. Notwithstanding the difficulty in measuring the costs and benefits of business incentives, there is little empirical evidence to support the proposition that business incentives affect industrial location decisions or economic growth. The next section reviews the empirical literature that tries to gauge the effects of incentives on plant location and state economic growth.

Empirical Studies

Over the years, analysts have used survey research and various statistical approaches to gauge the impact of business tax and financial incentives on industrial location decisions and state economic growth.

Survey Research

Through mail surveys and personal interviews, analysts can examine a number of qualitative variables that otherwise are difficult, if not impossible, to measure. And they can obtain bits of information about business investment decisions and uncover the significance of an array of factors that may not lend themselves to quantification. For example, Blair and Premus noted:

> *. . . quality of life is important to many firms, but it is a concept that is difficult to quantify. Even a factor such as 'access to markets' may be more meaningful to respondents than a rigorous, quantitative definition of the same concept.*[50]

There are many reasons why these surveys and interviews show mixed results regarding the impact of state business incentives on economic growth. Survey design, the types of industries chosen, advancement in industrial technology, the types of location factors selected and the timing of the survey can skew the results. For example, the surveys Due and Morgan reviewed were conducted in the 1940s and 1950s; as such, their studies do not account for technological changes and new market realities, nor do they reflect contemporary business decision-making regarding site selection.

However, some flaws are inherent in the survey questionnaires. Kieschnick noted that there can be ambiguity in the questions, in the responses received and in the measures used to interpret the responses.[72] For example, surveys variably might ask "What factors *were considered* in your location decision?" or "What factors *should* have been considered?" or "Did you consider factor 'X' in your decision?" Each question likely would result in a different response. Moreover, ranking factors as being of "no importance" or of "key importance" would not account for the magnitude of factors involved in actual location decisions.[73] And questions restricted to firms that have chosen a specific location, eliminate those that considered the area but ultimately rejected the location.[74] There also is little assurance that the appropriate business decision maker completed the survey.[75] Finally, as with any type of survey, there is an incentive for "strategic behavior" on the part of respondents who believe that their answers may influence the study's results.[76] The answers may be biased depending upon how the survey respondents hope to influence public policy toward business interests.[77]

Despite the flaws in some surveys, the results of a properly designed survey instrument could be useful in gaining a better perspective on state economic growth strategies. For example, most surveys suggest that traditional location factors such as markets, raw materials, labor and transportation are most important to firms' locational decisions. Advancements in industrial technology and new market realities might mean that state business incentives will become more important. Recall that the 1982 *Fortune* survey of factors important to companies considering plant location or relocation revealed that state financial inducements, ranked 23rd in 1976, moved up to 15th place five years later.[78] However, since most surveys have focused on tax incentives and not other state financial inducements, "few investigators have attempted to ascertain which incentives are viewed more favorably by industrial decision makers."[79] Surveys that identify which incentives are most favorable to business, which incentives are not of any value and which ones are of high value to business, but are not offered by states, could aid state economic development strategies.[80]

Statistical Approaches

Most of the early statistical studies found that business incentives had little influence on industrial location or state economic growth. For example, Clark Bloom[81] compared the growth in state and local tax collection during the periods 1939 to 1953 and 1947 to 1953 with the growth in manufacturing employment and capital expenditures. Simple correlation analysis showed no significance, although a small positive relationship did exist between taxes and manufacturing employment growth. But the study failed to fully account for other factors that could influence manufacturing employment growth. Moreover, the data included all taxes rather than just those that affected manufacturers.[82]

The earliest noteworthy study employing econometric techniques was conducted by Wilbur Thompson and John Matilla.[83] These researchers examined the annual employment growth of 29 manufacturing categories defined at the two-digit level by the Standard Industrial Classification System (SIC) from 1947 to 1954. They looked at interstate tax differentials paid by non-agricultural employees, along with taxes as a percentage of state personal income; changes in state population; changes in unionization; average industry wage rate; average educational level; industry investment and employment in 1947; and overall state manufacturing investment. With the exception of the apparel industry, state tax differentials had no significant influence on state economic growth.

Again, however, the study was lacking. Several scholars maintain that it suffered in the choice of tax measures and the lack of focus on industries capable of making significant choices in their location.[84] Furthermore, the study was hampered by a number of methodological problems, such as using different tax burden years to explain different manufacturing years, not partitioning tax burdens for different types of businesses and not determining whether state economic growth declined because of high tax burdens.[85]

In later years, however, other factors, including business climate, received more attention. While these early studies tried to measure the concept of "business climate," most concluded that business incentives were insignificant factors in state economic growth. Several that addressed the question of whether fiscal inducements other than taxes affect state economic growth also found that their role in promoting state economic growth was marginal. For example, Bridges' review of research on the effects of inducement programs concluded that financial inducements were secondary factors in a firm's location decision.[86] Kieschnick[87] found a weak statistical relationship between state economic growth and financial inducements, a conclusion echoed later by Steven Kale.[88]

Similarly, Dennis Carlton[89] found no significant link between business incentives and firm location decisions. Using Dun and Bradstreet data, he examined the impact of business incentives on the creation of new firms and the expansion of branch plants in three SIC industries: plastic products, electronic components and electronic communication equipment. To explain the birth rate and branch expansion of firms, Carlton compiled eight economic variables: wage rates (at the two-digit SIC level) collected from 1967 to 1973; state corporate and personal income tax rates for 1966 to 1975; effective property tax rates by Standard Metropolitan Statistical Area (SMSA); unemployment rates; utility costs (electricity and natural gas); proximity to markets and raw materials (agglomeration economies); technical expertise (measured by the number of engineers in the area); and a "business climate index" (state business incentive plans and policies).

Carlton found that wage rates and utility costs had a statistically significant effect on firm births and expansions. However, taxes and other business incentives had no significant effect on births and expansions for the industries under study. He concluded:

The models for both branch plants and single establishment firms tended to point toward several similar conclusions. First, for heavy energy using industries, the sensitivity of new location to energy costs is much higher than one would expect based on aggregate industry energy expenditure. Second, it is certainly not evident that taxes provide strong deterrents to locational activity, although the results were not sufficiently precise to rule out this possibility entirely. Third, for technologically sophisticated industries, local technical expertise can be important even for branch plants. Fourth, wages and agglomeration economies are likely to greatly influence new location. Finally, no evidence could be found to support the view that a favorable "business climate" alone can substantially stimulate new locational activity.[90]

In a subsequent study, Carlton[91] revised his model to examine the link between firm size and location. According to the author, decisions about branch location and size (number of employees) will predict "not only where new locations will occur but how much employment will be generated."[92] He found that while energy costs, existing concentrations of employment and available technical expertise affected the size of branch plants, state business incentives did not appear to have a major effect on branch plant size for any of the three industries, a finding similar to that of his 1979 study.

More recent research, however, is challenging the findings of these and other empirical studies.[93] Unlike earlier studies, where overall tax levels included both taxes and expenditures, contemporary studies are analyzing taxes and expenditures separately.

Thomas Romans and Ganti Subrahmanyan[94] used a simple model to determine the degree of tax progression in the tax structure and the extent to which income redistribution occurs through transfer payments. These authors speculated that higher progressive taxes and redistributing tax revenues through transfer payments drives out firms and high-income individuals, attracts low-income individuals and stagnates state economic growth. They found that the degree of progressivity in the personal tax structure and the amount of tax revenues flowing into transfer payments were negatively and significantly related to the growth in state personal income.[95] On the other hand, the level of personal taxes was unrelated to state economic growth. In fact, these authors found a positive relationship between state economic growth and the level of business taxes, and argued that it is due to the firms' willingness to pay extra for public services. However, others scholars have argued that this finding simply may reflect model misspecification.[96]

Like the traditional studies, most of the contemporary business climate studies by implication have included business incentives in their measure of the effectiveness of business climate on state economic growth. For example, Thomas Plaut and Joseph Pluta[97] analyzed pooled data for 48 states between 1967-72 and 1972-77 to determine the effectiveness of business climate factors on state economic growth, as measured by manufacturing growth.[98] Business climate was measured by taxes and expenditures, business costs and quality of life components, such as education, welfare, etc. States were ranked from 1 (best climate) to 48 (worst climate). Using three different measures of economic growth (including percentage change in employment, real value added and real capital stock), they found that business climate rankings and state economic growth did not have a strong relationship.[99]

Since the data indicated that factors other than business climate were important to state economic growth, the authors developed a model to test the effects of four groups of variables (accessibility to markets; cost and availability of factors of production, climate and environment; business climate; and state and local taxes and expenditures) on state economic growth. The results suggested that while traditional market factors, such as labor costs and climate, were more important in explaining differences in overall industrial expansion across states, business climate and taxation variables did influence manufacturing employment growth. Manufacturing employment growth was significantly related to labor and climate factors, and growth in real capital stock was related to the cost and availability of energy and land. After controlling for the effects of the other factors,[100] the group of business climate,[101] tax and expenditure variables was not significantly related to overall manufacturing growth, but was significantly related to state employment and capital stock growth.

Plaut and Pluta also found that corporate, sales and personal income taxes were least important to state industrial growth, but that there was a highly significant relationship between high local property taxes and all three measures of economic growth. The authors suggested that high local taxes might not be a deterrent to development if the benefits of the taxes, such as quality education and other services, are perceived by businesses to accrue locally. Indeed, they found high education expenditures positively related to employment growth. Thus, they concluded:

> While empirical support is therefore provided for the almost universal finding in the literature that individual state and local taxes have little effect on state industrial growth, our results suggest overall state and local tax effort is an important determinant of state employment growth. Even where business climate, tax, and expenditure variables were found to be significant determinants of regional growth, however, their role was still less important than that of traditional market factors (land and labor), newly emerging market factors (energy) and climate variables.[102]

Later, Timothy Bartik[103] looked at the relationship between various state characteristics and new branch plant locations.[104] Those state characteristics included effective corporate tax rate, business property tax rate, wages, unionization, education, population density, unemployment insurance, worker compensation, construction and energy costs, road miles and land area.

The results showed that unionization had a sizable statistical impact on new branch locations. The author cited that "a 10 percent increase in the percentage unionized of a state labor force is estimated to cause a 30-45 percent reduction in the number of new branch plants."[105] Bartik also found that the effective corporate and business property tax rates were negatively related to branch plant locations. He concluded:

The results . . . contradict the common view that state and local taxes and public services exert no influence on business location patterns. This has important implications for the incidence of state and local policies: as business location patterns change in response to taxes and services, land rents, local wages, and other prices will shift. These changes in business patterns put some limitation on the ability of states to redistribute income away from corporate stockholders both in state and out of state, and toward other state residents. On the other hand, the modest magnitude of this study's estimated tax effect means that states should not be fearful of becoming an economic wasteland due to slight increases in business taxation. In addition, state policy makers should not expect a supply-side miracle to result from tax incentives for new businesses.[106]

Similar results were found by Michael Wasylenko and Theresa McGuire.[107] The authors used an econometric model to examine the relationship between a state's business climate and employment growth rate. Using *County Business Patterns* data on employment trends between 1973 and 1980, they analyzed the aggregate percentage change of employment in six industries: manufacturing; transportation; wholesale trade; retail trade; finance, insurance and real estate; and services.[108]

The results showed that wages, energy costs, personal income taxation and increases in the overall tax level had a significant negative effect on employment growth. On the other hand, education spending as a percentage of income was related to employment growth in retail and finance industries. Welfare expenditures had an insignificant effect on employment growth. The authors suggested that even though factors beyond the control of policy-makers are major contributors to state employment growth, tax cuts still could affect the position of one state relative to another.

Overall, the findings on the effects of business incentives on state economic growth are mixed. Early studies suggested business incentives were not significant factors in state economic growth. However, their findings should be suspect since they suffered from a number of operational and methodological flaws. Those studies typically assumed taxes were adequate indicators of business incentives and rarely considered the impact of other fiscal inducements on business investment decisions. Similarly, those studies focused on plant location activities as indicators of state economic growth and paid little attention to employment or income growth.

Those early studies also relied heavily on simple correlation models to explain the relationship between incentives (defined as taxes) and state economic growth. The simple models did not take into account the fact that taxes are related to and are affected by other economic and fiscal factors. Consequently, the real impact of taxes, controlling for other fiscal factors, was unknown, and there was little or no information on the degree or extent to which these other factors affected state economic growth.[109] In addition, since early investigations focused on tax costs to businesses, tax benefits to businesses were never explored or properly weighed.

Finally, the measures of state tax burden used in these early studies failed to accurately reflect the actual tax cost to business and individuals. Past federal income tax deductions for businesses and individuals reduced the tax differential among states. More recent fiscal inducements may have only a negligible effect. These more recent studies also have shown that quality of life factors (e.g., education, recreation and other amenities) are significantly related to growth.

On the other hand, contemporary statistical studies, adjusting for the flaws of earlier works, have suggested that incentives in the form of tax breaks do influence state economic growth, although other federal tax reforms and the subsequent elimination of the deductibility component in federal corporate and individual income taxes might have a different impact on business investment decisions, although it is too early to estimate what that impact might be.

However, evidence of statistical relationships should not be equated with causation. There is *no* statistical evidence that business incentives actually create jobs, and there is *no* evidence that a loss of jobs or a transfer of jobs from one state to another is a direct result of business incentives. Businesses, to the extent they are mobile, will search for sites that offer the highest net advantage to them, taking into account the inherent features of the site. Different types of businesses are likely to put more emphasis on certain elements than on others.

Nevertheless, despite their inadequacies, these studies offer some useful primary insight into understanding the effectiveness of incentives. They suggest:

- Business incentives, as defined in these studies, *are not* the primary or sole influence on business location decision-making;

- Business incentives, relative to other factors, *do not* have a significant or primary effect on state employment growth;

- Business incentives *do* become more effective when all other variables are equal among competing sites within a region or sub-state area; and

- Business incentives *are* important in that they often are used as a component in business climate indices.

Interestingly, the overall findings on the significance of incentives are somewhat contradictory. Apparently, even though incentives have not proven to be important in effecting tangible improvement in state economic growth or attracting industry, they may still be important "psychological" or political weapons.

The Increase in Incentives — Why Has it Happened?

Despite the conflicting statistical evidence and uncertainty about the effects business incentives have on state economic growth, their availability as part of formal state programs has increased over the years. Keon Chi's inventory of state tax and financial incentive programs in 1988 shows that the number of states offering these sorts of incentives has drastically increased since 1976.[110] Since the literature suggests that the number of incentives a state offers does not reflect its relative importance to prospective businesses,[111] analysts now are turning their attention toward understanding the motivating factors behind the increase in business incentives.

There are several competing explanations for the rise in the availability and use of business incentives across the states. The traditional, and most obvious, explanation already mentioned in this report, is that their use stems from the states' need to expand economic growth either by increasing employment or attracting new industry. Much has been written about states "giving away the store" in order to attract businesses. As Enid Beaumont and Harold Hovey noted:

> *State and local officials commonly campaign on platforms that include strong planks on economic development — creating jobs and increasing economic activity. That state and local governments compete for economic development is not surprising. An expansion of jobs and population has many favorable impacts for those already located in a community. Land values, newspaper circulation, retail sales, and service business all increase when a state or local economy expands. In addition, political leaders pursuing economic development are fighting commonly perceived problems, such as the high unemployment rates experienced in the last recession.*[112]

Another explanation, however, is that the increasing use of incentives is a direct result of the business influence on state policymakers. For example, Sandra Kanter warned:

> *To a great extent, the business community has been able to influence legislative policy by playing one state against the other and convincing neighboring states of the need to keep their tax structures competitive with each other . . . until such time when there is better research and a system of pacts, it is incumbent for public interest lobbying groups to become more involved in state economic policy in order to counter the influence that business lobbyists have held in the past in forming policy.*[113]

To test this view, Margery Ambrosius interviewed 62 economic development agency officials and staff across five states and asked them who they perceived as having the most influence over their respective state's economic development policies: state elected officials, administrators or business lobbyists?[114] Ambrosius found that the economic development officials perceived elected officials, particularly governors, as being more instrumental in affecting state economic development policy.[115] However, she also discovered a difference in perspective between agency managers and technical staff. Persons at the managerial level viewed elected officials as the most influential actors in economic development policy, while individuals at the technical level saw business interests as most influential. She concluded:

> State economic development administrators appear to look first to elected officials, particularly the governor, for cues in undertaking administrative policymaking. The governor, through administrators' perceptions of gubernatorial importance, evidently plays an important role in shaping the implementation of state economic development policies. To a slightly lesser degree, these administrators turn to business interests for initiation of policies and thus, in all likelihood, for direction in implementation. According to this evidence and line of thought, business interests within the state play a strong secondary role in molding the final form of economic development policy.[116]

Recently, Ambrosius and Susan Welch[117] surveyed state legislators in Arizona, Georgia, Illinois, Massachusetts and Nebraska, to determine which interest group(s) these legislators viewed as most important. The legislators who responded[118] perceived business interests as more important than labor interests, in spite of their own party affiliation, ideology, tenure in the legislature and business occupation. Other analysts also have implicitly suggested that business groups play a major political role in economic decisions.[119] John Carroll, Mark Hyde and William Hudson tried to determine whether bureaucrats' and legislators' views on economic development policies converge as a result of various institutional changes in state government, including growing professionalism.[120] Their findings suggest that both legislators and bureaucrats tend to support business incentive policies, but that political pressures make legislators more sensitive than bureaucrats to political cues in their respective evaluations of economic development proposals. Political pressure to create jobs or prevent existing businesses from moving out may make legislators more sensitive to business interests, particularly if a large corporation publicly announces it will leave unless tax changes are implemented.

Yet another perspective on the increasing use of business incentives suggests that states share a fear of being "outbid." As explained by Rubin and Zorn:

> As regions, states and localities watch their neighbors attract jobs and economic activities the desire is to get a piece of the action. As more and more governmental units offer industrial location incentives to help tip business location decisions in their favor, support is lent to the belief that these incentives are necessary and that they significantly affect choices ... Policymakers are afraid that if they do not participate in the economic development bidding game (e.g., business incentives), their jurisdiction will lose jobs, economic stability and the appearance of vitality and robustness.[121]

One advocate of this "arms race" theory, Paul Peretz,[122] has argued that neither of the aforementioned notions — that politicians institute job growth programs for symbolic reasons or that business interests play a significant role in policy-making — is sound. He states that business incentives are more visible to the public and that their introduction generally will necessitate cutting services or raising personal taxes, actions which also are visible to the average voter. Moreover, he argues that the business interest theory fails to provide a good explanation for several reasons: (1) it is dependent on the plausibility of threats by business and the desperation of state economic development policy-makers, (2) it contradicts survey findings that incoming firms do not want to be seen as having been given special favors otherwise unavailable to other local businesses; (3) it ignores the constitutional and legislative constraints on offering new incentives; and (4) it underestimates the fact that very few incentives are the result of bargaining sessions with firms.

Peretz says that the growth in the use of incentives comes from the growth in economic development agencies, partly the result of federal subsidies for development efforts and the increasing tendency for corporations to look at wider areas and expect states and localities to provide site information. The increase in incentives, the author argues, can be explained by the "arms race" model, whereby states are forced into matching and beating incentives provided by other states because of their inability to act collectively.

Peretz's "arms race" explanation for the proliferation of financial incentives is supported by the recent research of Dennis O. Grady.[123] Grady reviewed the existing literature on the effects business incentives have on industrial location and job creation and reached the conclusion that incentives have little, if any, effect on either. In testing the competing explanations for the increasing use of incentives, Grady found that none of them held up. However, he argues "it appears that states enact incentives generally as a defensive measure against regional competitors."[124]

Although all of these studies view the business incentive competition among states as unhealthy, none has suggested that their use is likely to diminish. That being the case, if incentives are going to be used as policy instruments for economic development — regardless of their effectiveness, the motivating forces behind their use or the potentially harmful consequences — a more sound strategy for policy-makers might be to devise a method to determine which business incentives will be most advantageous and cost effective for their individual states.

Two studies focused on designing a framework in which states could achieve maximum benefits from business incentives. Rubin and Zorn[125] maintain that since business incentives have escalated regardless of economic soundness, state policy-makers should devise a method to achieve the most efficient use of these incentives. The authors argue that the use of location theory and comparative cost analysis provides a method to determine when and which incentives should be offered. They suggest that a state needs to estimate its uncontrollable costs, compare the advantages (or disadvantages) it has in terms of these costs, determine which industries give the state an advantage or competitive edge over other states and then offer or deny incentives accordingly. Rubin and Zorn stated:

Industrial expansion across states can be best explained by uncontrollable factors. On the whole, controllable factors (industrial location incentives) have little or no effect on firms' locational decisions. Controllable factors gain in significance only as the differentials in uncontrollables — transportation, energy, and labor costs — diminish.

... Presently, states and localities offer industrial location incentives indiscriminately. Indications are that many do not consider how they fare relative to their peers in terms of uncontrollable costs before offering location incentives. If a firm bases its decision on factors outside the control of the policy makers, it makes no sense to offer incentives which will not affect the decision... In other words, if a firm can be identified that would find the state or locality competitive in terms of uncontrollable costs, economic development efforts should be focused on attracting that firm.[126]

Other analysts have suggested that the tools of corporate finance might be used to determine the effectiveness of business incentives. David Rasmussen, Marc Bendick and Larry Ledebur,[127] used cost-benefit analysis to determine the type of business incentive that would minimize the public cost of a given level of financial assistance to business. These authors argue that state policy-makers should estimate the asset value of assistance[128] and the cost of state financial assistance to various levels of government, and then calculate the "cost-to-government/benefit-to-firm" ratio. The cost-to-government/benefit-to-firm ratio should allow policy-makers to maximize their benefits from business incentives while minimizing the cost to the public. They conclude that state development agencies should offer business incentive packages that are "tailored to a specific firm's requirements rather than a bundle of incentives that is invariant with the firm's situation."[129]

However, any meaningful comparative cost-benefit analyses will depend on an adequate understanding of some new forces at work in the economic development arena. Moreover, uncertainty and subjectivity affect state policy-makers' perceptions of costs and benefits. Indeed, as Ann Elder and Nancy Lind note:

...If a new plant locates in a community and expects to employ 1,400 individuals, presumably lowering the level of unemployment in the community, the untestable assumption is that the new workers will be hired from among the currently unemployed in the community. In order to evaluate the benefits of such a plant, moreover, officials must also make assumptions about continued employment levels and wage rates as well as levels of taxable profits that will be generated by the corporation.

These assumptions that are made — under conditions of uncertainty and subjectivity — will affect the decision makers' relative rank ordering of the alternatives and the respective weightings attached to each . . . .

The problem . . . is that economic development decisions are always made under conditions of uncertainty . . . the primary reason that cost/benefit ratios cannot be determined solely by mathematical formulas . . . .

In the decision-making process . . . the proximity of incurring costs or receiving benefits to the decision-making process influences the accuracy of the forecasts that influence the decision outcomes . . . [and] the perceived need for economic development influences the likelihood of decision makers succumbing to pressures to inflate expected benefits and deflate expected costs . . . .[130]

Concluding Observations

With the overall increase in industrial mobility, the emergence of high-technology firms and foreign investment and the changing nature of global competitive markets, the task of choosing the "right" or "best" economic development strategy should be a critical one for state policy-makers. Although the literature lacks consensus on the effectiveness of business incentives, it does offer some guidance for state policy-makers.

First, the type and mode of business investment will affect the impact of the incentive. As the literature suggests, capital-intensive manufacturing firms will be attracted by a different type of incentive than will service- or technical-oriented firms which are more labor-intensive. Similarly, plants that are expanding might require more land and tax concessions than start-up businesses that need venture capital. State policy-makers must determine what their goals are and which incentives can help them best achieve those goals.

Second, the concept of business climate, and what constitutes a favorable one, is changing. Early studies, focusing on domestic manufacturing firms, identified a favorable climate in terms of low land and labor costs and access to markets and raw materials. Later studies, focusing on high-technology and foreign firms, made "favorable climate" synonymous with the availability of skilled labor and high quality of life factors, such as universities and social and cultural amenities.

Competition among states for domestic and foreign investment is greater because of the changing business environment and the decreasing emphasis on production costs. This increased competition is accentuated by studies that rank states according to various business climate indices. Many analysts and practitioners already contend there are deficiencies in these studies.[131] However, state policy-makers need to be aware of the impact these climate studies might have on the direction and effectiveness of their economic development strategies. Reliance on the results of these rankings transforms business incentives from a policy tool designed to make a tangible improvement in the state's economic growth to one designed to improve the state's standing within someone's concept of a favorable business climate.

Whatever incentives the states use, guidelines must exist to keep policy-makers from feeling they need to "give away the store." In order to develop these guidelines, however, additional research and planning will be necessary. Future guidelines must incorporate quantifiable cost/benefit measures. These measures should include the amount of revenue lost because of the incentive, the impact the incentives have on competing services and expenditures, the number of jobs created by the incentive and the number of jobs lost to other states because of the incentives.

This review alludes to the complexity and non-quantifiable nature of the effects of business incentives. It includes uncertainty, subjectivity and other unknown and uncontrollable factors in the negotiation process that goes on between governments and businesses. That being the case, surveys of economic development officials as to how they award incentives would aid in determining the likely impact of guidelines on states and businesses. And further study of the negotiation process that ensues prior to the award of incentives might offer useful insight into a dilemma linked to the uncertainties of the competitive world economy — who should pay for what and how much in the case of default.

Notes

1. State economic growth, as used throughout this report, refers to increases in business activity (i.e., increases in plant locations, new firm births and expansion of existing businesses) and employment opportunities. While an expanded and diversified tax and revenue base is an important indicator of economic growth, it will not be explored in this report. Interestingly, many scholars implicitly assume that economic growth means that a state fosters a strong business climate or is competitive with other states. See Roger J. Vaughan, *State Taxation and Economic Development* (Washington, D.C.: Council of State Planning Agencies, 1979), pp. 3-18; John Gray and Dean Spina, "State and Local Industrial Location Incentives — A Well-Stocked Candy Store," *Journal of Corporation Law* 5 (Spring 1980): 521; William J. Barrett, VII, "Problems With State Aid to New or Expanding Business," *Southern California Law Review* 58 (1985): 1025-1026; Paul Peretz, "The Market for Incentives: Where Angels Fear to Tread?" *Policy Studies Review* 5 (February 1986): 624-633; and Dennis O. Grady, "State Economic Development Incentives: Why Do States Compete?" *State and Local Government Review* 19 (Fall 1987): 86-94.

2. Business incentives, as used throughout this report, refer to tax and financial subsidies, such as tax concessions, loans, revenue bonds, etc. Non-financial incentives such as educational facilities, cultural amenities, right-to-work laws or any other "quality of life" factors that promote a "good business climate," while noted, are not the focus of this report. In addition, this study, as is the case with most of the literature, does not make a clear distinction between business incentive programs and business incentive packages. The former refers to a set of policy instruments usually available to most firms through state economic development agencies. The latter is a set of incentives specifically tailored and packaged for a prospective firm. While studies have examined the effectiveness of business incentive programs on state economic growth, few have examined the effectiveness of business incentive packaging on economic growth. However, see H. Brinton Milward and Heidi Hosbach Newman, "State Incentive Packages and the Industrial Location Decision," paper presented at the Southern Political Science Association Meeting, Charlotte, N.C., November 5-8, 1987.

3. Will Myers, *Regional Growth: Interstate Tax Competition* (Washington, D.C.: Advisory Commission on Intergovernmental Relations, 1981), p. 5. Also see John E. Moes, "The Subsidization of Industry by Local Communities in the South," *Southern Economic Journal* 28 (October 1961): 187-193; John E. Moes, *Local Subsidies for Industry* (Chapel Hill, N.C.: University of North Carolina Press, 1962); James R. Rinehart, "Rates of Return on Municipal Subsidies to Industry," *Southern Economic Journal* 29 (April 1963): 297-306; James R. Rinehart, "Rates of Return on Municipal Subsidies to Industry: Reply," *Southern Economic Journal* 30 (April 1964):359-361; Ralph Gray, "Industrial Development Subsidies and Efficiency in Resource Allocation," *National Tax Journal* 17 (2) (June 1964): 164-172; and Gerald W. Samaza, "A Benefit Cost Analysis of a Regional Development Incentive: State Loans," *Journal of Regional Science* 10 (1970): 375-396. For a public official's advocacy of business incentives, see Ron Faucheux, "Louisiana's Incentive Package: What, Why, and How," speech delivered at the annual conference of the Public Affairs Research Council of Louisiana, Inc., Baton Rouge, La., March 14, 1985.

4. Will Myers, *Regional Growth: Interstate Tax Competition*, pp. 5-6. For criticism of Moes' "Subsidization of Industry," see Irving J. Goffman, "Local Subsidies for Industry: Comment," *Southern Economic Journal* 29 (October 1962): 111-114; and James H. Thompson, "Local Subsidies for Industry: Comment," *Southern Economic Journal* 29 (October 1962): 114-119. For criticism of Rinehart, see Lewis E. Hill, "Rates of Return on Municipal Subsidies to Industry: Comment," *Southern Economic Journal* 30 (April 1964): 358-359. Although the normative literature is replete with criticism concerning the states' use of business incentives, for an overall critique see Vaughan, *State Taxation and Economic Development*, in particular, Chapter 3; and Barrett, "Problems With State Aid to New or Expanding Businesses," pp. 1019-1050.

5. Keon Chi, *The States and Business Incentives: An Inventory of Tax and Financial Incentive Programs* (Lexington, Ky.: The Council of State Governments, 1989).

6. In 1640, the state of Massachusetts granted the first business incentive in the country. For a review of the long history of business incentives, see Sandra Kanter, "A History of State Business Subsidies," National Tax Association, *Proceedings of the Seventieth Annual Conference on Taxation* (Louisville, Ky.: 1977), pp. 147-55.

7. Peter W. Bernstein, "States are Going Down Industrial Policy Lane," *Fortune*, March 5, 1984, p. 112.

8. DeWitt John, *Shifting Responsibilities: Federalism in Economic Development* (Washington, D.C.: National Governors' Association, 1987), p. 69.

9. See James C. Cobb, "The Southern Business Climate: A Historical Perspective," *Forum for Applied Research and Public Policy* (Summer 1986), p. 95.

10. Gray and Spina, "State and Local Industrial Location Incentives," p. 521; Barrett, "Problems With State Aid to New or Expanding Business," p. 1020; and Jeffrey S. Luke, Curtis Ventriss, B.J. Reed and Christine M. Reed, *Managing Economic Development: A Guide to State and Local Leadership Strategies* (San Francisco, Calif.: Jossey-Bass Publishers, 1988), p. 195.

11. Barrett, "Problems With State Aid to New or Expanding Businesses," p. 1020. See also Susan A. MacManus, "Linking State Employment Training and Economic Development Programs: A 20-State Analysis," *Public Administration Review* (November/De-

12. Benjamin Bridges, Jr., "State and Local Inducements for Industry: Part I," *National Tax Journal* 18 (March 1965): 6.

13. Gray and Spina, "State and Local Industrial Location Incentives," p. 538; DeWitt John, *Shifting Responsibilities*, p. 69.

14. Mississippi's tax-exempt status allowed it to borrow monies at a much lower interest rate than private companies could, creating, in effect, a subsidy to prospective companies. Barrett, "Problems with State Aid to New or Expanding Businesses," p. 1020. Mississippi also granted companies locating in the state a five-year exemption from property taxes. DeWitt John, *Shifting Responsibilities*, p. 69.

15. Cobb, "The Southern Business Climate," p. 95.

16. For example, in 1956, the South Carolina legislature amended its alien properties laws to attract an English paper company. At about the same time, one South Carolina city reshaped its geography to provide a textile mill with a "tax-free island in the middle of the town." See Cobb, "The Southern Business Climate," pp. 95-96.

17. Bridges, "State and Local Inducements for Industry: Part I", pp. 1-14.

18. "The Second War Between the States: A Bitter Struggle for Jobs, Capital and People," *Business Week*, May 17, 1976, p. 92.

19. DeWitt John, *Shifting Responsibilities*, pp. 69-70.

20. Barry M. Rubin and C. Kurt Zorn, "Sensible State and Local Economic Development," *Public Administration Review* 45 (1985): 334.

21. R. Scott Fosler, *The New Economic Role of the American States: Strategies in a Competitive World Economy* (New York: Oxford University Press, 1988); and Luke, Ventriss, Reed and Reed, *Managing Economic Development*.

22. Fosler, *The New Economic Role of the American States*, pp. 3-5.

23. Keon Chi, *The States and Business Incentives*.

24. "War Between the States," *Newsweek*, March 30, 1988, p. 45.

25. Ibid., p. 44.

26. Grady, "State Economic Development Incentives," p. 87.

27. See Dennis Farney, "Nebraska, Hungry for Jobs, Grants Big Business Big Tax Breaks Despite Charges of Blackmail," *The Wall Street Journal*, June 23, 1987, pp. 66-67; and Don Wesely, "Myths and Realities of Economic Development Incentives, Who's Giving Away the Store," paper presented at the annual meeting of the National Conference of State Legislatures (Indianapolis, Ind., July 19, 1987).

28. M.J. Parks, "Washington State Scores a High-Tech Goal," *Business Week*, July 8, 1985, p. 31. See also, "Competition by States to Lure Firms Turns Into a Fierce Struggle," *The Wall Street Journal*, December 28, 1983.

29. The Volkswagen plant, in operation since 1978, received a $71 million incentive package from Pennsylvania. On November 21, 1987, it announced the closing of its Pennsylvania facility which eliminated 2,500 jobs. See, "Volkswagen to Close Only U.S. Plant Next Year," *The Washington Post*, November 21, 1987; and William Fulton, "VW in Pennsylvania: The Tale of the Rabbit That Got Away," *Governing* (November 1988), pp. 32-39.

30. Fulton, "VW in Pennsylvania: The Tale of the Rabbit That Got Away," p. 39.

31. Alfred Weber, *Theory of the Location of Industries* (Chicago: University of Chicago Press, 1929).

32. See Edgar M. Hoover, Jr., *The Location of Economic Activity* (New York: McGraw, 1948), in particular Chapter IV; Walter Isard, *Method of Regional Analysis: An Introduction to Regional Science* (New York: Wiley, 1960); David M. Smith, *Industrial Location: An Economic Geographical Analysis* (New York: Wiley, 1971).

33. Harold Hotelling, "Stability in Competition," *Economic Journal* 39 (1929): 41-57. For some of the shortcomings of the "least-cost" theory see Melvin L. Greenhut, *Plant Location in Theory and Practice: The Economics of Space* (Chapel Hill: University of North Carolina Press, 1956), pp. 255-263.

34. Greenhut, *Plant Location in Theory and Practice*, pp. 255-263.

35. August Losch, *The Economics of Location* (New Haven: Yale University Press, 1956). For an extensive review of Losch's theory, see Melvin Greenhut, *Plant Location in Theory and Practice*, Chapter II, sections I.1. and I.2.

36. Greenhut points out that Losch failed to "combine in his analysis an investigation of cost with an appraisal of locational interdependence" (p. 263). He also notes that a firm's profit varies by location. See Melvin Greenhut, *Plant Location in Theory and Practice*, pp. 263-269. In addition, the specific factors critical to a company's location decision vary by industry, since sales market and production inputs probably vary by industry. According to Michael Wasylenko:

> ...firms in industries using specialized technical workers may locate exclusively based on the supply of technical labor, while firms using less specialized labor may choose locations based on other criteria. Spatial clustering of firms may also occur if localization or agglomeration economies exist in certain industries. Thus, the conclusion about factors affecting firm location can be expected to vary by industry.

Wasylenko, "The Role of Taxes and Fiscal Incentives in the Location of Firms," in *Urban Government Finance: Emerging Issues*, ed. Roy Bahl, 20 (Beverly Hills, Calif.: Sage, 1981), pp. 157-158.

37. Greenhut, *Plant Location in Theory and Practice*, p. 256.

38. Walter Isard, *Location and the Space Economy* (New York: John Wiley, 1956), p. 264.

39. See G.B. Norcliffe, "A Theory of Manufacturing

Places," in *Locational Dynamics of Manufacturing Activity*, ed. Lyndhurst Collins and David F. Walker (New York: Wiley, 1975), pp. 14-58.

40. Ibid.

41. John P. Blair and Robert Premus, "Major Factors in Industrial Location: A Review," *Economic Development Quarterly*, 1 (February 1987): 73.

42. Ibid., pp. 74-75.

43. Ibid.

44. "Competition didn't count; productivity wasn't a factor in GE decision to close plant," (Louisville) *Courier Journal*, July 24, 1988.

45. Ibid.

46. Fulton, "VW in Pennsylvania: The Tale of the Rabbit That Got Away," p. 31.

47. "Washington State Scores a High-Tech Goal," p. 31.

48. See Fulton, "VW in Pennsylvania: The Tale of the Rabbit That Got Away," p. 31.

49. Cobb, "The Southern Business Climate," p. 98.

50. Blair and Premus, "Major Factors in Industrial Location: A Review," p. 76.

51. John F. Due, "Studies of State and Local Tax Incentives on Location of Industry," *National Tax Journal* 14 (June 1961): 163-173.

52. Ibid, p. 165.

53. William Morgan, "The Effects of State and Local Tax and Financial Inducements on Industrial Location," (Ph.D. Dissertation, University of Colorado, 1964).

54. Tables A and B show the results of Morgan's review of 17 survey studies and seven personal interview studies regarding the relative importance of location factors.

Table A
Relative Importance of Location Factors in 17 Surveys

Number of surveys in which a factor was described as being of:

Factor	Primary significance	Some significance	Little significance
Markets	16	1	0
Labor	10	7	0
Raw Materials	10	6	0
Transportation	7	10	0
Taxes	1	3	13
Financial Incentives	0	0	13

Source: William Morgan, "The Effects of State and Local Tax and Financial Inducements on Industrial Location" (Ph.D. dissertation, University of Colorado, 1964), Table 2.

Table B
Relative Importance of Location Factors in 7 Interview Studies

Number of interviews in which a factor was described as being of:

Factor	Primary significance	Some significance	Little significance
Markets	6	1	0
Labor	3	4	0
Raw Materials	3	4	0
Transportation	0	6	1
Taxes	0	0	7
Financial Incentives	0	0	7

Source: William Morgan, "The Effects of State and Local Tax and Financial Inducements on Industrial Location" (Ph.D. dissertation, University of Colorado, 1964), Table 4.

55. Stafford conducted in-depth interviews with primary decision makers in six multiplant firms about the location decisions of new plants. Using content analysis, the author first coded the respondents' statements according to a composite list of 14 location factors. The frequency of the statements indicated the relative importance of the location factors. See Howard Stafford, "The Anatomy of the Location Decision: Content Analysis of Case Studies," in *Spatial Perspectives in Industrial Organization and Decision Making*, ed. F.E. Hamilton (New York: Wiley, 1974), pp. 169-88. In addition, several early surveys of manufacturing firms in Michigan and Florida found that market-related factors, nontraditional factors (e.g., personal reasons, quality-of-life factors, etc.) played a major role in plant births, expansions and relocations. These studies found that taxes' and fiscal incentives' impact on industrial location decisions was minimal. See Eva Mueller, Arnold Wilken and Margaret Woods, *Location Decisions and Industrial Mobility in Michigan* (Ann Arbor, Mich.: Institute for Social Research, 1961); Eva Mueller and J.N. Morgan, "Locational decisions of manufacturers," *American Economic Review*, 52 (1962): 204-217; Melvin Greenhut and M.R. Goldberg, *Factors in the Location of Florida Industry* (Tallahassee: Florida State University Press, 1962); and Patricia A. Bradem and Susan R. Rideout, *Location Decision-Making in Export-Oriented Business and Industry* (Ann Arbor: Division of Research, Graduate School of Business Administration, University of Michigan, 1978).

56. Stafford, "The Anatomy of the Location Decision."

57. Roger W. Schmenner, "The Manufacturing Location Decision: Evidence from Cincinnati and New England," paper prepared for Economic Development Administration (Washington, DC.: U.S. Department of Commerce, 1978).

58. Ibid. Schmenner found that less than a third of

relocating plants moved into areas with lower property taxes, about one-half moved into similar taxing jurisdictions and about a fourth moved into areas with higher property taxes. In a subsequent study, Schmenner confirmed these findings for income tax rates. See Schmenner, *Making Business Location Decisions* (Englewood, N.J.: Prentice Hall, 1982).

59. Roger W. Schmenner, *The Location Decisions of Large Multiplant Companies* (Washington, D.C.: U.S. Department of Housing and Urban Development, 1980); and Schmenner, *Making Business Location Decisions*. In extensive interviews with 80 key location decisions, Schmenner found that none of these decisionmakers said taxes were significant in their location decisions. Interestingly, some interviewees felt that business incentives would create poor community relations and that no controls existed to prevent a state from rescinding on incentives. Schmenner notes, "In the long run both the company and the town will benefit, it is thought, if the company makes a point of paying its fair share." See Schmenner, *Making Business Location Decisions*, pp. 46-47.

60. Schmenner, *Making Business Location Decisions*, p. 51.

61. John S. Hekman, "Survey of Location Decisions in the South," *Economic Review*, (June 1982): 6-19. Hekman conducted mail surveys and telephone interviews on business incentives from 204 predominantly branch operations of multiplant firms in North Carolina, South Carolina and Virginia between 1977 and 1981. These executives were asked to rank 19 location factors and 12 quality-of-life factors on a scale of one to five.

62. Fortune Inc, *Facility Location Decisions* (New York: Fortune Inc., 1977).

63. Fortune Market Research Survey, "Why Corporate America Moves Where" (New York: Times Inc., 1982), p. 9.

64. "Building and Site Selection," *Inc Magazine* (Boston, Mass., 1980).

65. Mark L. Goldstein, "Choosing the Right Site," *Industry Week*, April 15, 1985, pp. 57-60.

66. Michael Kieschnick, *Taxes and Growth: Business Incentives and Economic Development* (Washington, D.C.: Council of State Planning Agencies, 1981).

67. John Rees, "Regional Industrial Shift in the U.S. and the Internal Generation of Manufacturing Growth Centers in the Southwest," in *Interregional Movement and Regional Growth*, ed. W. Wheaton, Coupe Paper 2 (Washington, D.C.: Urban Institute, 1979); Roger W. Schmenner, *The Location Decision of Large Multiplant Companies*; and Blair and Premus, *Major Factors in Industrial Location*, pp. 79-81.

68. Robert Premus, *Location of High Technology Firms and Regional Economic Development*, U.S. Congress Joint Economic Committee (Washington, D.C.: Joint Committee Print, 1982).

69. Out of 1,750 firms surveyed, 691 firms (or 39.5 percent) responded.

70. Premus predicted that the New England, mid East and far West areas would not maintain their high tech growth because of major problems with labor costs and availability, taxes, congestion, housing costs and lack of land for expansion. On the other hand, the Midwest, South, Mountain and Plain areas were expected to receive the largest percentage increase in new high tech firms during the coming years. The findings are consistent with Rees' 1979 survey of high tech firms. See Rees, "Regional Industrial Shift in the U.S."

71. Mamoru Yoshida, "The Investment Decision-Making Process," in *Japanese Direct Manufacturing Investment in the United States* (New York: Praeger Publishers, 1987), pp. 41-74. The Japanese executives were asked to rank the importance of 12 locational factors on a four-point scale ranging from very important to not important.

72. Michael Kieschnick, *Taxes and Growth*, pp. 52-53.

73. Ibid.
74. Ibid.
75. Ibid.
76. Ibid.
77. Ibid.

78. Fortune Market Research Survey, "Why Corporate American Moves Where," p. 9.

79. Steven R. Kale, "U.S. Industrial Development Incentives and Manufacturing Growth During the 1970's," *Growth and Change*, 15 (January 1984): 26-34.

80. For survey studies that attempt to address this issue, see Industrial Development Research Council, *The Industrial Facility Planner's View of Special Incentives* (Atlanta: Industrial Development Research Council, 1977); Richard A. Duvall, "Industry, States Rate Incentives, Assistance Programs," *Industrial Development*, November/December 1968, pp. 26-30; and V.P. Apilado, "Public Administration of Financial Incentives in Industrial Plant Location: Industrial Aid Bonds," *Papers in Public Administration*, 26 (Arizona State University, Institute of Public Administration, 1973). While several states have assessed industry attitude toward their respective states, none has apparently determined the effectiveness of business incentives on its economic growth. See Yankelovich, Skelly and White, Inc., "A Survey of Business Executive Attitudes Toward Wisconsin as a Business Location," prepared for the Wisconsin Department of Development, April 1984; Keith L. Borders and Jerry Johnson, *Economic Development: A Question of Strategy* (Oklahoma: Economic Policy Interim Study, August 1985); Wisconsin Department of Revenue, *Corporate Tax Climate: A Comparison of 16 States* (Madison: Wisconsin Legislative Council, 1982); and Runyon, Kersteen, Ouellette and Co., *Maine Business Climate Satisfaction Survey* (Portland, Maine: Author, 1986).

81. Clark C. Bloom, *State and Local Tax Differentials* (Iowa City: Bureau of Business Research, State University of Iowa, 1955).

82. John Due, "Studies of State-Local Tax Incentives

Bartik, Timothy et al. "Saturn and State Economic Development." *Forum for Applied Research and Public Policy*, Spring 1987, p. 37.

Bartsch, Charles. "Michigan: Reaching for Recovery." *Commentary*, Fall 1986, pp. 8-12.

Beaumont, Enid F., and Hovey, Harold A. "State, Local, and Federal Economic Development Policies: New Federal Patterns, Chaos, or What?" *Public Administration Review* 45 (March/April 1985): 327-332.

Beckman, Bruce A., and Praire, Michael W. "A Guide to Obtaining Required Regulatory Approval for New Industrial Facilities in California." *San Diego Law Review* 17 (August 1980): 979-1019.

Berman, Norton L. "Trends in Economic Development: Sending a Message to Development Professionals." *Commentary*, Spring 1986, pp. 14-16.

Bird, Richard M. "Tax Subsidy Policies for Regional Development." *National Tax Journal* 19 (2) (1966): 113-124.

Blair, John; Fichtenbaum, Rudy; and Swaney, James. "The Market for Jobs: Location Decisions and the Competition for Economic Development." *Urban Affairs Quarterly*, September 1984, pp. 64-77.

Bond, E.W., and Samuelson, L. "Tax Holidays as Signals." *American Economic Review* 76 (September 1986): 820-826.

Bridges, Benjamin, Jr. "State and Local Inducements for Industry: Part I." *National Tax Journal* 18 (March 1965): 1-14.

———. "State and Local Inducements for Industry: Part II." *National Tax Journal* 18 (June 1965): 175-192.

Carroll, John; Hyde, Mark; and Hudson, William. "Economic Development Policy: Why Rhode Islanders Rejected the Greenhouse Compact." *State Government* 58 (3) (1985): 110-112.

———. "State-Level Perspectives on Industrial Policy: The Views of Legislators and Bureaucrats." *Economic Development Quarterly* 1 (November 1987): 333-340.

Cobb, James C. "The Southern Business Climate: A Historical Perspective." *Forum For Applied Research and Public Policy* 1 (Summer 1986): 94-100.

Conway, McKinley. "A Technology Forecast for Development Strategies: The Next 30 Years, 1986-2016." *Economic Development Review* 4 (Summer 1986): 33-40.

Dewar, Margaret. "Development Analysis Confronts Politics: Industrial Policy on Minnesota's Iron Range." *APA Journal*, Summer 1986, pp. 290-298.

Dubnick, Mel. "American States and the Industrial Policy Debate." *Policy Studies Review* 4 (1984): 22-27.

Dye, Thomas R. "Taxing, Spending, and Economic Growth in the American States." *Journal of Politics* 42 (November 1980): 1081-1107.

Elder, Ann H.; and Lind, Nancy S. "The Implications of Uncertainty in Economic Development: The Case of Diamond Star Motors." *Economic Development Quarterly* 1 (February 1987): 30-40.

Fleagle, Kenneth R. "Patterns of Success in Luring Industry." *Arizona Review of Business and Public Administration* 13 (11) (November 1964): 1-5.

Foster, John L. "Regionalism and Innovation in the American States." *The Journal of Politics* 40 (1978): 179-187.

Fox, William F., and Neel, Warren C. "Saturn: The Tennessee Lessons." *Forum for Applied Research and Public Policy* 1 (Spring 1987): 7-16.

Friedland, R. "The Politics of Profit and the Geography of Growth." *Urban Affairs Quarterly* 19 (September 1983): 41-54.

Goffman, Irving J. "Local Subsidies for Industry: Comment." *Southern Economic Journal* 29 (October 1962): 111-114.

Grady, Dennis O. "State Economic Development Incentives: Why Do States Compete?" *State and Local Government Review* 19 (Fall 1987): 86-94.

Gray, John and Spina, Dean. "State and Local Industrial Location Incentives — A Well Stocked Candy Store." *Journal of Corporation Law* 5 (Spring 1980): 517-687.

Gray, Ralph. "Industrial Development Subsidies and Efficiency in Resource Allocation." *National Tax Journal* 17 (2) (June 1964): 164-172.

Grossman, Ilene K. "Meeting the Challenge (The New Industrial Revolution)." *Public Law Forum* 4 (Spring 1985): 419-426.

Harline, Osmond L. "How States Compete for New Industry." *Utah Economic and Business Review* 24 (8) (August 1964): 1-11.

Harrison, Bennett, and Kanter, Sandra. "The Great State Robbery." *Working Papers for a New Society*, Spring 1976, pp. 54-66.

———. "The Political Economy of State Job Creation Business Incentives." *Journal of American Institute of Planners* 44 (November 1978): 424-435.

Hayden, F. Gregory; Kruse, Douglas C.; and William, Steve C. "Industrial Policy at the State Level in the United States." *Journal of Economic Issues* 19 (1985): 383-96.

Henderson, William L., and Fleagle, R. Kenneth. "A Novel Industrial Financing Technique." *Arizona Review of Business and Public Administration* 13 (9) (September 1964): 4-8.

Hyde, Mark; Hudson, William; and Carroll, John. "Business and State Economic Development." *Western Political Quarterly* 41 (March 1988): 181-191.

Jacobs, Jerry. "Corporate Subsidies From the 50 States." *Business and Society Review*, Spring 1980, pp. 47-50.

Kaiser, Susan E. et al. "Providing Financial Incentives for Industry to Remain at its Urban Location." *Urban Law Annual* 21 (1981): 334-342.

Krutilla, John V. "Criteria for Evaluating Regional Development Programs." *American Economic Review* 45 (2) (May 1955): 120-132.

Levitt, Arthur Jr. "Industrial Policy: Slogan or Solution?" *Harvard Business Review*, March/April 1981, pp. 6-8.

Lind, Nancy S. and Elder, Ann H. "Who Pays? Who Benefits? The Case of the Incentive Package Offered to the Diamond Star Automotive Plant." *Government Finance Review*, 2 (6), December 1986: 19-23.

Liner, C.D. "State Economic Development Policies: A Review of Four Reports." *Popular Government* 52 (4) (Spring 1987): 61-63.

McManus, Susan A. "Linking State Employment and Training and Economic Development Programs: A 20-State Analysis." *Public Administration Review*, November/December 1986: pp. 640-650.

McNamee, Stephen J. "Dupont - State Relations." *Social Problems*, February 1987, pp. 1-17.

Merlin, Matthew R. "Industrial Development Bonds at 50: A Golden Anniversary Review." *Economic Development Quarterly* 1 (November 1987): 391-407.

Milward, H. Brinton. "Can a Governor Determine the Location of Private Industry?" *Kentucky Economic Review and Perspective* 10 (Summer 1986): 3-4.

Mitchell, John N. "Municipal Industrial Aid Bonds." *Municipal Finance* 33 (4) (May 1961): 163-168.

Moes, John E. "The Subsidization of Industry by Local Communities in the South." *Southern Economic Journal* 28 (October 1961): 187-193.

Moore, Michael C. "Idaho Industrial Development Bonds." *Idaho Law Review* 20 (Winter 1984): 39-62.

Netzer, Dick. "State Tax Policy and Economic Development: What Should Governors Do When Economists Tell Them That Nothing Works?" *New York Affairs*. (3) (1986): 19-36.

Peretz, Paul. "The Market For Incentives: Where Angels Fear To Tread?" *Policy Studies Review* 5 (February 1986): 624-633.

Rees, John. "Industrial Innovation: Linking New Production Methods and Regional Growth Rates." *Economic Development Commentary*, Spring 1986, pp. 17-21.

———. "Technological Change and Regional Shifts in American Manufacturing." *Professional Geographer*. 31 (1) (February 1979): 45-54.

Reich, Robert B. "An Industrial Policy of the Right." *The Public Interest*, Fall 1983, pp. 3-17.

Rosenfeld, Stuart. "Southern Strategies for Economic Development." *Forum For Applied Research and Public Policy* 2 (Summer 1987): 48-55.

Ross, D. William. "Tax Exemption in Louisiana as a Device for Encouraging Industrial Development." *Southwestern Social Science Quarterly*, June 1953, pp. 14-22.

Schwartz, Gail Garfield. "The New Realities of Economic Development." *Southern Review of Public Administration* 6 (Winter 1983): 390-405.

Smith, Paula V. "State and Local Policy: Efforts in Missouri (The New Industrial Revolution)." *Public Law Forum* 4 (Spring 1985): 427-432.

Solomon, Lewis, and Stern, Janet. "Enterprise Zones, Tax Incentives and the Regulation of Inner Cities: A Study of Supply-Side Policy Making." *Detroit College of Law Review*, Fall 1981, pp. 797-836.

Stephenson, Susan C., and Hewitt, Roger S. "State Tax Revenues Under Competition." *National Tax Journal* 37 (1984): 95-101.

———. "Strategies for States in Fiscal Competition." *National Tax Journal* 38 (1985): 219-226.

Sternberg, Ernest. "A Practitioner's Classification of Economic Policy Instruments, with Some Inspiration from Political Economy." *Economic Development Quarterly* 1 (May 1987): 149-161.

Stober, William J. and Falk, Lawrence H. "Industrial Development Bonds as a Subsidy to Industry." *National Tax Journal* 22 (1969): 232-243.

Surrey, Stanley S. "Tax Incentives as a Device for Implementing Government Policy: A Comparison with Direct Government Expenditures." *Harvard Law Review* 83 (January 1970): 705-738.

Thornburgh, Richard. "State Strategies and Incentives for Economic Development." *Journal of Law and Commerce* 4 (Summer 1984): 1-17.

Vaughan, Roger J. "State Tax Incentives: How Effective Are They?" *Commentary*, January 8, 1980, pp. 3-5.

Watkins, Alfred J. "Good Business Climates: The Second War Between the States." *Dissent* 27 (Fall 1980): 476-485.

Weinstein, Bernard L. "Tax Incentives for Growth." *Society* 14 (3) (1977): 73-75.

3. Documents, Reports, Unpublished Papers, Theses, Etc.

Ambrosius, Margery M. "State Economic Development Administrators: Innovators, Representatives, or Neutrals?" Paper delivered at the 1987 annual meeting of the American Political Science Association, Washington, D.C., September 1984.

Beier-Solberg, Ann. *State and Local Taxes and Economic Growth—Review of the Literature.* Madison: Wisconsin Department of Revenue, Division of Research and Analysis, September 1984.

Borders, Keith L., and Johnson, Jerry. *Economic Development: A Question of Strategy.* Oklahoma: Midwest Political Science Association, Chicago, Ill., April 10-12, 1986.

Brace, Paul, and Baumann, Philip. "Markets Versus the Polity: The Politics of State Economic Growth." Paper delivered at the 1986 annual meeting of the Midwest Political Science Association, Chicago, Ill., April 10-12, 1986.

California Department of Commerce. *Facts: The Californias.* Sacramento: 1986.

California Department of Commerce. Office of Business Development. *The Californias: Business Incentives.* Sacramento: February 1988.

California Department of Commerce. Office of Economic Research. *The Californias: Environmental Permits*. Sacramento: 1988.

California Economic Development Corporation. *1987 Annual Report*. Sacramento: 1987.

Dubnick, Mel. "Industry Policy in the American States: The Invisible Side of the Debate." Paper presented at the 1983 annual meeting of the Southern Political Science Association, Birmingham, Ala., November 3-5, 1983.

Dubnick, Mel, and Holt, L. "National Industrial Policy and States: Roles and Options." Paper presented at National Conference on Public Administration, Denver, Colo., April 9-11, 1984.

Faucheux, Ron. "Louisiana's Incentive Package: What, Why, and How." Speech delivered at the annual conference of the Public Affairs Research Council of Louisiana, Inc., Baton Rouge, La., March 14, 1985.

Florida Department of Commerce. Division of Economic Development. *Project Cornerstone*. Tallahassee: October 1988 (revised).

Floyd, Joe S., Jr. "Federal, State and Local Government Programs for Financing Industrial Development." *Proceedings of the 55th Annual Conference on Taxation*. National Tax Association, Harrisburg, Pa., 1962.

———. "State and Local Financing for Industrial Development." *Proceedings of the 56th Annual Conference on Taxation*. National Tax Association, Harrisburg, Pa., 1963.

Floyd, Joe S., Jr., and Hodges, Luther H., Jr. *Financing Industrial Growth: Private and Public Sources of Long Term Capital for Industry*, Research Paper No. 10, Chapel Hill, N.C.: School of Business Administration, University of North Carolina, May 1962.

Fosler, R. Scott. "State Economic Development: What Have We Learned and Where Are We Going?" Paper prepared for the National Conference of State Legislatures, Conference on State Economic Development and Competitiveness Policies, Boston, Mass., June 1987.

———. "Strategies for State Economic Development." Presentation to Leadership Conference, Maryland House of Delegates, Annapolis, Md., December 14, 1987.

Grady, Dennis O. "The Evolution of State Economic Development Policy." Paper presented at the 1986 annual meeting of the Midwest Political Science Association, Chicago, Ill., April 10-12, 1986.

———. "Marketing the State: The Governor's Role in State Trade Promotion." Paper presented at the 1985 annual meeting of the Midwest Political Science Association, Chicago, Ill., April 17-20, 1985.

Grossman, Ilene K. "Economic Development Incentives: Guidelines for Decision-Making." A report prepared for the Midwestern Legislative Conference Business Development Task Force, January 1988.

———. "Economic Development: States Battle to Attract Jobs — An Overview of the Business Tax Incentive Issue." Paper presented at the 1987 annual meeting of the Midwestern Legislative Conference of The Council of State Governments, August 1987.

Hansen, Susan B. "State Industrial Policy: The Case of Pennsylvania." Paper presented at the 1985 annual meeting of the Southwest Political Science Association, Houston, Texas, March 20-23, 1985.

———. "State Perspectives on Economic Development: Priorities and Outcomes." Paper presented at the 1986 annual meeting of the Midwest Political Science Association, Chicago, Ill., April 10-12, 1986.

Harrison, Bennett. *The Economic Development of Massachusetts*. Prepared for the Joint Committee on Commerce and Labor of the Massachusetts General Court, November 1974.

Hawaii Department of Business and Economic Development. *Envision Hawaii: Strategic Plan: Business and Industry Development, Marketing and Promotion*. Honolulu: January 1988.

Heare, Jerry. *Attracting New Industry*. Austin, Texas: Texas Industrial Commission, 1980.

Illinois Department of Commerce and Community Affairs. *Sell Illinois: A Strategy for the Present, a Commitment to the Future*. Springfield: 1987.

———. *State of Illinois Five Year Economic Development Strategy*. Springfield: March 1988.

Illinois Legislative Council. "Catalog of Industrial Development Financial Incentive Programs." November 1, 1982.

———. "Tax Incentive in Aid of Urban Redevelopment." Memo to Senator Robert W. McCarthy, 1968.

Indiana Economic Development Council, Inc. *Looking Back: The Update of Indiana's Economic Development Plan - An Evaluation of Progress to Date, In Step With the Future*. Volume I. Indianapolis: September 1987.

———. *Looking Forward: The Update of Indiana's Strategic Economic Development Plan - Strategies for the Future*. Volume II. Indianapolis: June 1988.

Indiana State Chamber of Commerce. *In Step With the Future - Indiana's Strategic Economic Development Plan*. Indianapolis: 1983.

Industrial Development Research Council. *The Industrial Facility Planner's View of Special Incentives*. Atlanta: Industrial Development Research Council, 1977.

Institute for Urban Studies and Community Service. *Economic Development Target Industry Study*. Charlotte, N.C.: University of North Carolina at Charlotte, 1982.

Iowa Department of Economic Development. "Directions for Iowa's Economic Future: Executive Summary." *New Opportunities for Iowa: Strategic Planning Recommendations for Economic Development*. Des Moines: March 1987.

Johnson, William A. "Industrial Tax Exemptions: Sound Investment or Foolish Giveaway?" *Proceedings of the 55th Annual Conference on Taxation*. National Tax Association, Harrisburg, Pa., 1962.

Kanter, Sandra. "A History of State Business Subsidies." *Proceedings of the 70th Annual Conference on Taxation*, National Tax Association, Louisville, Ky., 1977.

Kentucky Economic Development Cabinet. Reported *Foreign Direct Industrial Investment in Kentucky*. Frankfort: April 1989.

Liner, Blaine, and Ledebur, Larry. "Foreign Direct Investment in the United States: A Governor's Guide." Paper prepared for the 79th Annual Meeting of the National Governors' Association, July 26-28, 1987.

Liner, Charles D. *Business Taxation and Economic Development in North Carolina*. Chapel Hill, N.C.: Institute of Government/University of North Carolina at Chapel Hill, March 1983.

Lock, Bill. *Economic Development: A Review of State Actions*. Lincoln: Nebraska Legislative Council, Legislative Research Division, December 1986.

McIntyre, Robert S., and Tipps, Dean C. *The Failure of Corporate Tax Incentives: A Study of Growing Loopholes and Lagging Investment*. Washington, D.C.: Citizens for Tax Justice, January 1985.

Maine Economic Development Strategy Task Force, Maine Development Foundation. *Establishing the Maine Advantage: An Economic Development Strategy for the State of Maine*. Augusta: October 1987.

Marketing Wisconsin Task Force. Final Report to James T. Flynn, Lt. Governor. Madison: Wisconsin Department of Development, 1983.

Milward, H. Brinton and Newman, Heidi Hosbach. "State Incentives Packages and the Industrial Location Decision." Paper presented at the 1987 annual meeting of the Southern Political Science Association Meeting, Charlotte, N.C., November 5-8, 1987.

Minnesota. *The Report of the Governor's Commission on the Economic Future of Minnesota*. St. Paul: 1987.

Minnesota House of Representatives Research Department. *Minnesota's Economic Development Programs: A Guide for Legislators*. St. Paul: 1984.

———. *Constitutionality of Minnesota Electric Corp. v. Tully and Westinghouse Electric Corp. v. Tully and Credits: A Legal Analysis*. St. Paul: 1985.

New Jersey Governor's Office of Policy and Planning. *State Policy Options for High Technology Promotion*. Trenton, N.J.: 1981.

New York Department of Commerce. *Tax Incentives and Financing Assistance for Industrial Location*. Research Bulletin, No. 52, 1983.

North Carolina Department of Commerce. *North Carolina: The Better Business Climate. Economic Advantages of Warehousing and Distribution*. Raleigh: 1987.

Ohio Department of Development. *Ohio's Thomas Edison Program*. Columbus: August 1985.

Ohio Department of Development. *Resource Ohio: Financial Assistance, Employment and Training, Technical Assistance, Applied Technology and Research*. Columbus, Ohio, no date.

Pennsylvania Economic Development Partnership Office. "Investment in Pennsylvania's Future: The Keystone for Economic Growth." *An Executive Summary of Pennsylvania's Economic Development Partnership Office*. Harrisburg: January 1988.

Radatz, Clark G. *Wisconsin's Strategy to Spur Economic Development: A Summary of the 1983 Special Session of the Wisconsin Legislature*. Wisconsin: Legislative Reference Bureau, 1984.

Rappa, John. *Economic Development Programs*. Hartford: Connecticut Office of Legislative Research, 1984.

———. "Industrial Development Assistance." Memo to Joint Committee on Legislative Affairs, Connecticut General Assembly. Hartford: Office of Legislative Research.

Rhode Island Planning Program. Providence: March 1986.

Sanzone, John G. "Issues in Economic Development." Background papers prepared for the Second Annual Northern California Economic Development Conference, Chico, Calif. University Center for Economic Development and Planning, 1987.

Skoro, C.L. "Ranking of State Climates: An Evaluation of Their Usefulness in Forecasting." Paper prepared for the Pacific Northwest Economic Conference, Seattle, Wash.: April 1987.

South Dakota Governor's Office of Economic Development. *Statewide Action Program for Economic Development*. Pierre: February 1988.

SRI International Center for Economic Competitiveness. *New Seeds for Nebraska: Strategies for Building the Next Economy*, 1987.

———. *Profiles of Key Industries in Pennsylvania: Competing in the Global Marketplace*. January 1988.

Steinbach, Carol. "Economic Development in the States: There's a New Look Coming." Paper prepared for the Council of State Planning Agencies, 1980.

Suzman, Cedric J., and Heslin, J. Alexander. "An Evaluation of Current Trends in Foreign Direct Investment in the U.S." Southern Center for International Studies Conference Paper, March 29, 1985.

Texas Industrial Commission. *Financing Industrial Facilities in Texas*. Austin: May 1980.

Texas Strategic Economic Development Commission. *A Blueprint for Tomorrow's Texas*. Austin: May 1988.

Vass, Tom. *Industrial Recruitment and the Path of North Carolina's Economic Development to the Year 2000*. Raleigh: North Carolina Department of Labor, 1982.

Washington, Office of the Governor. *Washington Economic Development Agenda: Priorities and Strategies.* Olympia: January 1988.

Wesely, Don. "Myths And Realities Of Economic Development Incentives: Who's Giving Away The Store?" Paper presented at the 1987 annual meeting of the National Conference of State Legislatures, Indianapolis, Ind., July 19, 1987.

Wisconsin Department of Development. *Final Report: Governor's Advisory Committee on Business Incentives.* Madison: September 1987.

Wisconsin Department of Development. *Models of State Entrepreneurial Development Programs.* Madison: 1987.

Wisconsin Department of Revenue, Division of Research and Analysis. *Corporate Tax Climate: A Comparison of 16 States.* Madison: 1982.

Wisconsin Strategic Development Commission. *Phase I - The Mark of Progress.* Madison: 1985.

_____. *Wisconsin Strategic Development Commission: The Final Report.* Madison: 1986.

4. *Newspaper Articles, Magazines and Miscellaneous Publications*

Arnold, W.F. "California Bolsters Plans to Keep Companies In-State." *Electronic Business*, January 1983, p. 152.

Battle, B. "Secrecy Yields Saturn Plant for Tennessee." *National Real Estate Investor*, October 1985, p. 9.

"Bidding for Saturn." *Fortune*, April 1, 1985, p. 9.

Bernstein, Peter W. "States are Going Down the Industrial Policy Lane." *Fortune*, March 5, 1984, pp. 112-113.

Brennan, M. "Tennessee Sheds Its Hick Image for High Tech." *Electronic Business*, August 15, 1985, p. 132.

"Calif. Senate Okays Bill on Tax Incentive for Foreign Firms." *Platt's Oilgram*, June 27, 1985, p. 3.

"Company Meets Community: Long Courtship, Happy Marriage." *Nation's Business*, November 1984, pp. 52A-52P.

"Competition didn't count: productivity wasn't a factor in GE decision to close plant." (Louisville) *Courier Journal*, 24 July 1988.

Cook, B. "The Garden State: Planting for High-Tech Growth." *Electronic Business*, September 1, 1985, p. 136.

Curtis, C.E. "Stalking Smokestacks." *Forbes*, August 1, 1983, p. 48.

"Debate Over Cost of Getting Car Plant." *The Pantagraph*, December 14, 1985.

"Do Venture Capitalists Really Need a Tax Break?" *Business Week*, April 8, 1985, p. 100.

Eckhouse, J. "The Siliconization of Scott's Valley: Too Much, Too Soon?" *Electronic Business*, October 1, 1984, pp. 55-56.

Engardino, P., and Edid, M. "Why A Little Detroit Could Rise in Tennessee." *Business Week*, August 12, 1985, p. 21.

Farney, Dennis. "Nebraska, Hungry for Jobs, Grants Big Business Big Tax Breaks Despite Charges of 'Blackmail.'" *Wall Street Journal*, 23 June 1987, pp. 66-67.

"Financial Assistance for Industry." *Industrial Development*, January/February 1985, p. 52.

"Financial Assistance for Industry," *Site Selection Handbook*, October 1985, p. 900.

Fitch, E. "Saturn Ad Has a Stellar Copywriter." *Advertising Age*, September 19, 1985, p. 56.

"Forecast 1986: Hot Spots." *Buildings*, January 1986, pp. 48-53.

"Fort Howard Will Spend $1 Billion on Georgia Mill." *Paper Trade Journal* 169 (September 1985): 16.

Frazier, D. "New Jersey Counts High Tech Instead of Smokestacks." *Electronic Business*, February 1983, p. 134.

Fulton, William. "VW in Pennsylvania: The Tale of the Rabbit That Got Away." *Governing*, November 1988, pp. 32-39.

Fyffe, David E. "IRB's Can Be Cost Effective." *Industrial Development* 153 (January/February 1984): 30-33.

Garfield, B. "Spring Hill Saturnalia." *Advertising Age*, August 18, 1986, p. 72.

"General Motors: Rings of Saturn." *Economist*, August 3, 1985, pp. 61-62.

Gooding, Edwin C. "New War Between the States, Parts I-IV," *New England Business Review*, Federal Reserve Bank of Boston, Boston, Mass., October 1963, December 1963, July 1964 and October 1964.

Gregg, G. "Helping Themselves: Pennsylvania Program Aids Fledgling Firms." *Barron's*, October 8, 1984, p. 62.

Hartfield, H. "Politics Triumphs in State Bid Wars for Plum Projects." *ENR*, September 10, 1987, pp. 18-21.

"Harvester is Expected to Close Indiana Unit, Switch to Ohio." *Wall Street Journal*, 27 September 1982, p. 8, col. 3.

"Harvester Lenders Skeptical of New Plant to Restructure 4.2 Billion Credit Accord." *Wall Street Journal*, 3 August 1982, p. 8, col. 1.

"Hayssen Shifts All Production to SC." *Plastics World*, January 1986, p. 13.

Herbers, John. "Industrial Flight in Minnesota." *New York Times* April 12, 1983, pp. D1 and D19.

"The High Tech Rennaisance in Southern California." *Business Week*, September 17, 1984, pp. 142-144.

"Hollywood's Dollars Go On Location." *Los Angeles Times*, 8 January 1984, sec. I, p. 1, col. 1.

Inaba, M. "Japan Cos. Spurn California for East, North." *Electronic News*, August 18, 1986, supp. 11.

"Kalamazoo May Be Home to GM's Saturn Plant." *National Journal* 17 (29) (1985): 1663.

Kern, R. "Smokestack California." *Industrial Development*, May/June 1986, pp. 26-27.

Kieschnick, Michael. "States Can Increase Capital Flows." *Ways and Means* (Spring 1987).

King, J.L. "Columbus Means Business to the Japanese." *Automotive News*, June 16, 1986, p. D16.

"Knauf Fiber Glass Chooses Lanett, Ala. for Site of New Fiber Glass Insulation Plant." *Industrial Development*, March/April 1987, pp. 26-27.

Knee, R. "The Lure of Los Angeles." *American Shipper*, May 1986, p. 56.

Kotlowitz, Alex, and Buss, Dale D. "Localities Giveaways to Lure Corporations Cause Growing Outcry; Tax Breaks, Other Incentives Viewed as Wasteful by Unions and Taxpayers." *Wall Street Journal*, 4 September 1986, p. 1-21.

Krebs, Emilie. "State, Local Officials Led Bid for Plant." *The Pantagraph*, 18 April 1986.

Krebs, M. "GM Confirms Tennessee as Saturn Site." *Automotive News*, August 5, 1985, p. 2.

———. "Michigan Offers More Saturn Deals." *Automotive News*, September 9, 1985, p. 64.

Lancaster. "Competition by States to Lure Firms Turns Into a Fierce Struggle." *Wall Street Journal*, 28 December 1983, p. 1, col.6.

Lane, M.B. "Connecticut Searches for the High-Tech Connection." *Electronic Business*, August 15, 1984, pp. 131-132.

Leonhardt, D. "Environmental Aspects of the Site Location Decision." *Industrial Development*, July/August 1984, pp. 23-27.

Lichtenstein, Eugene. "Higher and Higher Go the Bids for Industry." *Fortune* April 1964, pp. 118-121.

Liston, Linda. "States Legislate Generous Remedies for Adverse Business Environment." *Industrial Development* 138 (6) (November/December 1969): 5-8.

Loftus, Tom. "Indiana Auto-Plant Incentives to Go Beyond $55 Million From Tax Funds." (Louisville) *Courier-Journal*, 4 December 1986, sec. B2.

"La. Petroleum Groups Back Extension of Tax Incentive Law." *Platt's Oilgram*, April 22, 1987, p. 5.

Lowenstein, Roger. "NBC Opts to Stay in New York City Due to Tax Breaks." *Wall Street Journal*, 9 December 1987, sec. 2G, p. 18, col. 1

Luxenburg, S. "States for Sale." *Madison Avenue*, October 1984, pp. 81-84.

Lyne, J. "The Geo-Political Index: How Government Influences the Facility Planning Process." *Site Selection Handbook*, October 1985, pp. 872-874.

———. "Washington Wins RCA/Sharp from Oregon in Last-Second Ploy." *Industrial Development*, September/October 1985, pp. 28-29.

McCosh, D.F. "Detroit Should Start Trying to Fathom California." *Wards Auto World*, October 1983, p. 17.

McElroy, J. "Tennessee: There Ain't No Place GM'd Rather Be." *Automotive Industries*, September 1985, p. 24.

"Mack Trucks Relocates Assembly Plant in South Carolina." *Industrial Development*, May/June 1986, pp. 26-27.

Mallory, A. "New Breed Stalks Adventure in Investment." *Grand Rapids Press*, 3 June 1984, sec. E1, col. 1.

Markos, A.J. "IRB's Play Vital Role in Rhode Island's Revitalization." *United States Banker*, September 1982, p. 75.

Miller, S.W. "Site Selection: New Game, New Rules." *Railway Age*, June 1985, pp. 36-37.

Miyauchi, Takeo. "The Man Who Lured Toyota to Kentucky." *Economic Eye*, March 1987, pp. 23-27.

Moore, J.W. "Corporate Kidnapping." *National Journal* 19 (24) (June 13, 1987): 1518-1521.

"Munster Pins Hope on Grant." (Hammond, Ind.) *Times*, 28 December 1983, sec. 4, col. 3.

National Governors' Association. "Survey of State Linkages Between Employment, Training and Economic Development." *Labor News* (July 1985a), pp. 15-17.

"N.D. Lawmakers Pass Incentives Package Pegged on WTI Price." *Platt's Oilgram*, April 22, 1987, p. 6.

"North and South Battle for Industry." *Marketing and Media Decisions*, June 1981, pp. 66-69.

O'Connor, M. "High-Tech Core Areas Thriving in California and Along East Coast." *Site Selection Handbook*, June 1986, pp. 556-558.

Olson, Walter. "Industrial Policy from the Grass Roots." *Wall Street Journal*, 12 June 1984, p. 30.

Parks, M.J. "Washington State Scores a High-Tech Goal." *Business Week*, July 8, 1985, p. 31.

"PBS Settles on Alexandria as New Home." *Broadcasting*, May 13, 1985, pp. 101-102.

"PBS Settles into its New Home." *Broadcasting*, March 17, 1986, pp. 107-109.

"Pharmaceutical Companies are Flocking to Connecticut." *Chemical Marketing Reporter*, July 22, 1985, p. 24.

Phelan, M. "Mixed Welcome: Fuji, Isuzu Dedicate New Indiana Home Site." *Wards Auto World*, January 1987, p. 55.

Pilcher, Dan. "Economic Development: Old Term Has New Meaning." *State Legislature* 12 (7) (1986): 18-21.

———. "Michigan: The Road to Recovery." *State Legislatures*, Winter 1983, pp. 17-21.

"Plant Bait Battle." *Chemical Week*, August 17, 1963.

"Reverse Investment: Is South Korea Next Big U.S. Plant Builder?" *Industry Week*, March 18, 1985, p. 28.

Rozen, M. "State Programs Lure High Tech Companies." *Dun's Business Month*, March 1985, p. 93.

"Rust Bowl a Magnet for Investments." *Los Angeles Times*, 19 November 1984, sec. I p. 1, col. 1.

Scanlon, J.R. "Site Selection and Design for the Growth Industries." *Industrial Development*, March/April 1984, pp. 26-29.

Schellhardt, Timothy. "War Among the States for Jobs and Business Becomes Ever Fiercer." *The Wall Street Journal*, 14 February 1984, sec. 1A.

Schroeder, M. "State Ads Entice Foreign Investors." *International Advertiser*, October 1986, p. 32.

"The Second War Between the States: A Bitter Struggle for Jobs, Capital, and People." *Business Week*, May 17, 1976, pp. 92-114.

Serafin, R., and Snyder, J. "How 2 States Pitched Saturn." *Advertising Age*, August 5, 1985, p. 4

"$70 Million Aids Research Facility Lands in Tucson." *Industrial Development*, July/August 1985, pp. 28-29.

"Silicon Valley Loses One as Hi-Tech Firm Picks Cascade." *Grand Rapids (Michigan) Press*, 15 July 1984, sec. G1, col. 1.

Smart, Lucien E. "Midwestern State Plots Economic Strategy - (Indiana)." *Public Utilities Fortnightly*, August 2, 1984, pp. 6-7.

Smith, D.C. "The Missing Q in Saturn Q&A." *Wards Auto World*, September 1985, p. 33.

"The Spoils of Saturn." *Fortune*, September 2, 1985, p. 9.

"State Incentives for Pollution Control." *Site Selection Handbook*, October 1986.

"State of Indiana Makes a Pitch for High Tech." *National Real Estate Investor*, September 1984, p. 36.

"States Battle It Out." *Management Review*, June 1985, p. 4.

Stein, Herbert. "Don't Fall for Industrial Policy." *Fortune*, November 14, 1983, pp. 64-86.

Sutton, Horace. "Sunbelt vs. Frostbelt: A Second Civil War?" *Saturday Review* 5 (14) (April 15, 1978): 28-37.

Swasy, Alice. "Dollars, Doubts Line Japan's 'Auto Alley'." *Lexington Herald-Leader*, 28 June 1987, Sec. A1.

"Tax Incentives for Industry: Other Laws." *Industrial Development*, January/February 1983.

"Tennessee's Perspective on the GM Saturn Plant Location Decision." *Industrial Development*, March/April 1986, pp. 22.

"Two Towns Fight to Keep Harvester Plants, Know Only One Will Remain Open," *Wall Street Journal*, 8 September 1982, p. 35, col. 4.

U.S. Department of Commerce. "State Assistance, Incentives and Services to Industry". *Economic Development* 10 (February 1973).

"Using Tax-Exempts to Build Business." *Business Week* December 14, 1963.

"Volkswagen to Close Only U.S. Plant Next Year." *The Washington Post*, November 21, 1987.

"The War Between the States." *TWA Ambassador*, December 1981, pp. 23-26.

Weinstein, Bernard L., and Gross, Howard T. "What Counts Most in the Race for Development." *State Legislatures*. May/June 1988, pp. 22-24.

"Why Oregon Suddenly Looks Good to High Tech Companies (Elimination of State Unitary Tax)." *Business Week*, November 5, 1984, p. 138.

Witten, M. "To Move Or Not To Move Is Only the First Question: With Thousands of Development Agencies Eager to Attract High-Tech Industries, Electronics Companies Can Afford to Be Choosy." *Electronic Business*, May 15, 1983, p. 20.

"Would Industrial Policy Help Small Business?" *Business Week*, February 6, 1984, pp. 72-73.

B. Location Theories and Processes

1. *Books and Monographs*

Allaman, Peter M., and Birch, David L. *Components of Employment Change for States by Industry Group, 1970-72*. Cambridge, Mass.: Joint Center for Urban Studies of Massachusetts Institute of Technology and Harvard University, September, 1975.

Creamer, David. *Changing Location of Manufacturing Employment, Part I*. New York: National Industrial Conference Board, Studies in Business Economics, 83, 1963.

Fisher, James S.; Hanink, Dean M.; and Wheeler, James O. *Industrial Locational Analysis: A Bibliography, 1966-1979*. Athens: University of Georgia, Department of Geography, 1979.

Fortune, Inc. *Facility Location Decisions*. New York: Fortune Inc. 1977.

Fuchs, Victor R. *Changes in the Location of Manufacturing in the United States Since 1929*. New Haven, Conn.: Yale University Press, 1962.

Greenhut, Melvin L. *Plant Location in Theory and Practice: The Economics of Space*. Chapel Hill, N.C.: University of North Carolina Press, 1956.

Greenhut, Melvin L., and Goldberg, Marshall R. *Factors in the Location of Florida Industry*. Tallahassee: Florida State University, 1962.

Harris, Curtis C., and Hopkins, Frank E. *Locational Analysis*. Lexington, Mass.: Lexington Books, 1972.

Hoover, Edgar M. *The Location of Economic Activity*. New York: McGraw-Hill, 1948.

Hunker, Henry L., and Wright, Alfred Jr. *Factors of Industrial Location in Ohio*. Columbus: Ohio State University, Bureau of Economic Research.

Isard, Walter. *Location and the Space Economy*. New York: John Wiley, 1956.

———. *Method of Regional Analysis: Introduction to Regional Science*. New York: John Wiley, 1960.

Losch, August. *The Economics of Location*. New Haven, Conn.: Yale University Press, 1956.

Mandell, L. *Industrial Location Decisions: Detroit Compared with Atlanta and Chicago*. New York: Praeger, 1975.

Miller, E.W. *Manufacturing: A Study of Industrial Location*. University Park: Pennsylvania State University Press, 1977.

Moriarty, B.M. *Industrial Location and Community Development.* Chapel Hill, N.C.: University of North Carolina Press, 1980.

Norcliffe, G.B. "A Theory of Manufacturing Places." In *Locational Dynamics of Manufacturing Activity,* pp. 14-58. Edited by Lyndhurst Collins and David F. Walker. New York: Wiley, 1975.

Rees, John, and Stafford, Howard. "High Technology Location and Regional Development: The Theoretical Base." In *Technology, Innovation, and Regional Economic Development.* Washington, D.C.: U.S. Congress, Office of Technology Assessment, 1984.

Smith, David M. *Industrial Location: An Economic Geographical Analysis.* New York: Wiley, 1971.

Stafford, Howard A. *Principles of Industrial Facility Location.* Atlanta: Conway Publications, 1980.

Stevens, Benjamin H. and Brackett, Carolyn A. *Industrial Location: A Review and Bibliography of Theoretical, Empirical and Case Studies.* Philadelphia: Regional Science Research Institute, 1967.

Stone, Donald B. *Industrial Location in Metropolitan Areas.* New York: Praeger, 1976.

Struyk, Raymond J., and James, Franklin. *Intermetropolitan Industrial Location: The Pattern and Process of Change in Four Metropolitan Areas.* Lexington, Mass.: Lexington Books, 1975.

Weber, Alfred. *Theory of the Location of Industries.* Translated by C.J. Friedrich. Chicago: University of Chicago Press, 1929.

Wheat, Leonard F. *Regional Growth and Industrial Location.* Lexington, Mass.: Lexington Books, 1973.

Wolman, Hal. *Components of Employment Change in Local Economies.* Washington, D.C.: The Urban Institute, 1978.

Yaseen, Leonard C. *Plant Location.* New York: American Research Council, 1960.

Yoshida, Mamoru. "The Investment Decision-Making Process." In *Japanese Direct Manufacturing Investment in the United States,* pp. 41-74. New York: Praeger, 1987.

2. *Articles*

Armington, C.; Harris C.; and Odle, M. "Formation and Growth in High-Technology Firms: A Regional Assessment." In *Technology, Innovation, and Regional Economic Development.* Washington, D.C.: U.S. Congress, Office of Technology Assessment, 1984.

Blair, John P. and Premus, Robert. "Major Factors in Industrial Location: A Review." *Economic Development Quarterly* 1 (February 1987): 72-85.

Cameron, Helen A. "Property Taxation as a Location Factor." *Bulletin of Business Research* 44 (1969).

Carlton, Dennis. "Location Decisions of Manufacturing Firms." Report 7728. Chicago: Center for Mathematical Studies in Business and Economics, University of Chicago, 1977.

Chait, Stephen J. "Branch Plants: How to Pursue Them and Keep Them." *Economic Development Review* 5 (Winter 1987): 36-39.

Chinitz, Benjamin, and Vernon, Raymond. "Changing Forces in Industrial Relocation." *Harvard Business Review* 38 (1) (January 1960): 26-36.

Crandall, Robert W. "The Transformation of U.S. Manufacturing." *Industrial Relations* 25 (Spring 1986): 118-130.

Dean, Robert D., and Carroll, Thomas M. "Plant Location Under Uncertainty." *Land Economics* (November 1977).

Duerkson, Christopher J. "Industrial Plant Location: Do Environmental Controls Inhibit Development?" *Economic Development Commentary,* Winter 1985, pp. 17-21.

Dunn, Edgar S., Jr. "The Market Potential Concept and the Analysis of Location." *Regional Sciences Association Papers and Proceedings,* 1956.

Erickson, R., and Wasylenko, M. "Firm Relocation and Site Selection in Suburban Municipalities." *Journal of Urban Economics* 8 (1980): 69-85.

Fulton, Maurice. "New Factors in Plant Location." *Harvard Business Review* 49 (3) (May/June 1971): 4-17.

Gage, John C., Jr. "Changing Site Selection Requirements." *Economic Development Review* 5 (Winter 1987): 33-35.

Harris, Chauncy, D. "The Market as a Factor in the Location of Industry in the United States." *Annals of the Association of American Geographers.* 44 (44) (December 1954): 315-348.

Hekman, John and Smith, Alan. "Behind the Sunbelt's Growth: Industrial Decentralization." *Economic Review,* March 1982, pp. 4-13.

Hotelling, Harold. "Stability in Competition." *Economic Journal* 39 (1929): 41-57.

Johnson, E. "New Rules to Use in Site Location Today." *Dun's Business Month,* November 1986, pp. 73-90.

McMillan, T.E. "Why Manufacturers Choose Plant Locations vs. Determinants of Plant Location" *Land Economics,* August 1965, pp. 232-38.

Markusen, Ann. "High-Tech Plants and Jobs: What Really Lures Them." *Commentary,* Fall 1986, pp. 3-7.

Mueller, Eva, and Morgan, J.N. "Locational Decisions of Manufacturers." *American Economic Review* 52 (1962): 204-217.

North, DC. "Locational Theory and Regional Economic Growth." *Journal of Political Economy,* June 1955.

Rubin, Barry M. and Zorn, C. Kurt. "Site Location: State Taxes, Fees and Labor Costs." *Distribution,* May 1983, pp. 77-87.

Schmenner, Roger W. "Location Decisions of Firms, Implications of Public Policy." *Economic Development Commentary,* Winter 1981, pp. 3-7.

―――. "Look Beyond the Obvious in Plant Location," *Harvard Business Review,* January/February 1979, pp. 126-32.

Schneider, Mark. "Suburban Fiscal Disparities and the Location Decisions of Firms." *American Journal of Political Science* 29 (August 1985): 587-605.

Singer, J.F., and Sarb, P.J. "A Model for Location and Development Decision Making." *American Journal of Small Business* 7 (October-December 1982): 3-11.

Singhvi, S.S. "A Quantitative Approach to Site Selection." *Management Review*, April 1987, pp. 47-52.

Steinnes, Donald N. "Causality and Intraurban Location." *Journal of Urban Economics* 4 (No. 1 1977): 59-79.

3. Documents, Reports, Unpublished Papers, Theses, Etc.

Bradem, Patricia A., and Rideout, Susan R. *Location Decision-Making in Export-Oriented Business and Industry*. Ann Arbor, Mich.: Division of Research, Graduate School of Business Administration, University of Michigan, 1978.

Economic Development Administration. *Survey of Industrial Location Determinants*. Washington, D.C.: U.S. Department of Commerce, 1971.

McGuire, Theresa J. "Essays on Firm Location in a Metropolitan Area," Ph.D. thesis submitted to Princeton University, June 1983.

Mueller, Eva; Wilken, Arnold; and Woods, Margaret. *Location Decisions and Industrial Mobility in Michigan*. Ann Arbor, Mich.: Institute for Social Research, University of Michigan, 1961.

Schmenner, Roger W. "The Location Decision of Large Multiplant Companies." Washington, D.C.: U.S. Department of Housing and Urban Development, 1980.

4. Newspaper Articles, Magazines and Miscellaneous Publications

Ady, Robert M. "Shifting Factors in Plant Location." *Industrial Development*, November/December 1981, pp. 13-17.

Allen, B. "Corpus Christi - A Regional Site Selection Influence (Saturn and Homeport Projects)." *Industrial Development*, September/October 1985, pp. 24-25.

"Atlanta Picked Best Area to Locate Business." *Nation's Business*, November 1984, p. 41.

Baum, L. et al. "Does It Pay to Move the Corporate Headquarters?" *Business Week*, September 7, 1987, pp. 68-69.

Ciandella, D. "TRW Locates Near Market, Employees." *Industrial Development*, March/April 1985, pp. 366-367.

Conway, McKinley. "Terrorism: Growing Factor in Location Decisions." *Site Selection Handbook* 31 (August 1986): 952-956.

Cooper, T.C. and O'Connor, M. "Technical Support for Corporate Facility Planners Grows in Scope and Depth." *Site Selection Handbook* 30 (April 1985): 306-307.

Donovan, D.J. "Twelve Key Questions for Site Selection Decision Makers." *Industrial Development*, July/August 1982, pp. 12-15.

Ellenis, M. "Six Major Trends Affecting Site Selection Decisions to the Year 2000." *Dun's Business Month*, November 1983, pp. 116-130.

Feinberg, P. "Site Selection: Wooing the Corporate Customer." *Institutional Investor*, May 1981, p. 187.

"Frostbelt States Become Attractive to Relocating Firms." *National Real Estate Investor*, June 1984, p. 26.

Gabe, V.D. "An Outline of a Corporate Site Location Procedure." *Industrial Development*, September/October 1983, pp. 23-25.

Goldstein, Mark L. "Choosing the Right Site." *Industry Week*, April 15, 1985, pp. 57-60.

"Guide to Distribution & Site Location Decisions." *Distribution*, September 1986, p. 20.

Hack, G.D. "The Plant Location Decision Making Process." *Industrial Development*, September/October 1984, pp. 31-33.

Harding, Charles F. "Company Politics in Plant Location." *Industrial Development*, September/October 1982, pp. 19-20.

———. "New Plant Location Strategies." *Dun's Business Month*, November 1984, pp. 111-112.

Harrington, L.H. "Anatomy of a Site Selection." *Traffic Management*, December 1983, pp. 62-64.

Holub, K.M. and Levin, J.E. "Strategic Waste Management Consideration in Siting Industrial Facilities." *Industrial Development*, November/December 1984, pp. 4-13.

Lamberson, M. "Industrial Development Bonds Can Tip the Balance in Site Selection." *Industrial Development*, January/February 1977, pp. 16-18.

Lewis, M. "What It Takes Now to Lure Businesses to Your Area." *Nation's Business*, May 1984, pp. 36A-36H.

Martin, T.J. "Urban Sites Proving Attractive for New Industry." *National Real Estate Investor*, February 1983, p. 30.

"Money Isn't Everything in Choosing a Spot." *Nation's Business*, November 1984, p. 41.

"Now Energy is What Counts in War Between States." *Business Week*, October 26, 1981, p. 166.

O'Connor, M. "Federal Cuts, Regulatory Changes Fail to Reduce Services for Site-Seekers." *Site Selection Handbook*, April 1987, p. 64-66.

"Plant Site Preferences and Industry and Factors of Selections." *Business Week Research Report*, 1959.

"Site Makes Right." *Dun's Business Month*, October 1985, Supp D.

"Site Selection: Questions to Consider When Relocating." *Traffic Management*, June 1985, pp. 27-28.

"Three Perspectives on Industry's Migration." *Industry Week*, January 6, 1986, pp. 31-32.

"To Move Or Not To Move?" *Fortune*, December 8, 1986, p. 32.

Wardrep, B.N. "Factors Which Play Major Roles in Location Decisions." *Industrial Development*, July/August 1985, pp. 8-12.

C. Empirical and Business Climate Studies

1. Books and Monographs

Advisory Commission on Intergovernmental Relations. *The Michigan Single Business Tax*, M-114. Washington, D.C.: Advisory Commission on Intergovernmental Relations, March 1978.

———. *State-Local Taxation and Industrial Location*. Washington, D.C.: Advisory Commission on Intergovernmental Relations, 1981.

Bloom, Clark C. *State and Local Tax Differentials*. Iowa City: Bureau of Business Research, State University of Iowa, 1955.

Campbell, A. "State and Local Taxes, Expenditures and Economic Development." In *State and Local Taxes on Business*. Princeton, N.J.: Tax Institute of America, 1965.

Carlton, Dennis. "Why New Firms Locate Where They Do: An Econometric Model." In *Interregional Movement and Regional Growth*, pp. 13-50. Edited by W. Wheaton. Washington, D.C.: The Urban Institute, 1979.

Dewar, Margaret E. *The Usefulness of Industrial Revenue Bond Programs for State Economic Development: Some Evidence from Massachusetts*. Cambridge, Mass.: Joint Center for Urban Studies of the Massachusetts Institute of Technology and Harvard University, March 1980.

Fisschel, W. "Fiscal and Environmental Consideration in the Location of Firms in Suburban Communities." In *Fiscal Zoning and Land-Use Controls*. Edited by E.S. Mills and W.E. Oates. Lexington, Mass.: DC. Heath, 1975.

Harrison, Bennett. "Regional Restructuring and Good Business Climate: the Economic Transformation of New England Since World War II." In *Sunbelt Snowbelt: Urban Development and Regional Restructuring*, pp. 48-98. Edited by Larry Sowers and William Tubb. New York: Oxford University Press, 1984.

Luger, Michael I. *The States and High Tech Development: The Case of North Carolina*. Durham, N.C.: Duke University Institute of Policy Sciences and Public Affairs, September 1984.

Malinowski, Z.S., and Kinnard, W.N., Jr. *Personal Factors Influencing Small Manufacturing Plant Locations*. Storrs, Conn.: Small Business Administration, Management Research Grant Program, School of Business Administration, Connecticut State University.

Mandell, L. *Industrial Location Decisions: Detroit Compared with Atlanta and Chicago*. New York: Praeger, 1975.

Margolis, Wendy and Felton, Denise. *Locating Industry in Arkansas: York-Hannover, A Case Study in Public Incentives*. Little Rock, Ark.: The Winthrop Rockefeller Foundation, 1984.

Morgan, William E. *Taxes and the Location of Industry*. Boulder, Colo.: University of Colorado Press, 1967.

Oakland, William. "Local Taxes and Intra-Urban Industrial Location: A Survey." In *Metropolitan Financing and Growth Management*. Edited by G. Break. Madison: University of Wisconsin, 1978.

Premus, Robert. *Location of High Technology Firms and Regional Economic Development*, Washington, D.C.: U.S. Congress Joint Economic Committee Print, 1982.

Rees, John. "Regional Industrial Shift in the U.S. and the Internal Generation of Manufacturing Growth Centers in the Southwest." In *Interregional Movement and Regional Growth*. Edited by W. Wheaton. Washington, D.C.: The Urban Institute, 1979.

Sack, S. "State and Local Finance and Economic Development." In *State and Local Taxes on Business*. Princeton, N.J.: Tax Institute of America, 1965.

Schmenner, Roger W. *Making Business Location Decisions*. Englewood Cliffs, N.J.: Prentice-Hall, 1982.

———. *Summary of Findings: The Manufacturing Location Decision: Evidence from Cincinnati and New England*. Cambridge, Mass.: Harvard Business School, 1978.

Stafford, Howard. "The Anatomy of the Location Decision: Content Analysis of Case Studies." In *Spatial Perspectives on Industrial Organization and Decision-Making*, pp. 169-185. Edited by F.E. Hamilton. New York: John Wiley, 1974.

Strasma, D. *State and Local Taxation of Industry: Some Comparisons*. Boston: Federal Reserve Bank of Boston, 1959.

Survey Research Center. *Industrial Mobility in Michigan*. Ann Arbor, Mich.: University of Michigan Press, 1950.

Suzman, Cedric J. et al. *The Costs and Benefits of Foreign Direct Investment From a State Perspective*. North Carolina: Southern Center for International Studies, August 1982.

Thompson, Wilbur, and Matilla, John M. *An Econometric Model of Postwar State Industrial Development*. Detroit: Wayne State University Press, 1959.

Wasylenko, Michael. "The Location of Firms: The Role of Taxes and Fiscal Incentives." In *Urban Government Finance*, pp. 155-190. Edited by R. Bahl. Beverly Hills, Calif.: Sage Publications, 1981.

Whitman, Edmund S., and Schmidt, W. James. *Plant Relocation: A Case History of a Move*. New York: American Management Association, 1966.

2. Articles

Adams, Jack E.; Lewison, Dale M.; and Rucks, Conway, T. "Public Industrial Location Inducement: Snowbelt-Sunbelt Preference." *Review of Regional Economics and Business*, October 1979, pp. 33-40.

Bahl, Roy W., and Shellhammer, Kenneth L. "Evaluating the State Business Tax Structure: An Application of Input-Output Analysis." *National Tax Journal* 23 (2) (1969): 203-216.

Bartik, Timothy. "Business Location Decisions in the United States: Estimates of the Effects of Unionization, Taxes and Other Characteristics of States." *Journal of Business and Economic Statistics* 3 (January 1985): 14-22.

Bergin, Thomas P., and Eagan, William F. "Economic Growth and Community Facilities." *Municipal Finance* 33 (4) (May 1961): 146-150.

Bozeman, Barry, and Bozeman, Lisle, J. "Manufacturing Firms' View of Government Activity and Commitment to Site: Implications for Business Retention Policy." *Policy Studies Review* 6 (3) (February 1987): 538-553.

Carlton, Dennis. "The Location and Employment Choices of New Firms: An Econometric Model with Discrete and Continuous Endogenous Variables." *The Review of Economics and Statistics* 65 (1983): 440-449.

Charney, A.H. "Intraurban Manufacturing Location Decisions and Local Tax Differentials." *Journal of Urban Economics* 14 (September 1983): 184-205.

Chernotsky, H.I. "Selecting U.S. Sites: A Case Study of German and Japanese Firms." *Management International Review* 23 (2) 1983: 45-55.

Conway, Carol. "Foreign Direct Investment in the South: A Review of Data and Recruitment." *Southern International Perspectives*, June 1985, pp. 1-26.

Dahl, David, and Gene, Samuel. "The Impact of State and Local Taxes on Economic Growth." *Regional Science Perspectives* 10 (2) (1980).

Digby, Michael F. "Evaluating State Industrial Development Programs." *Southern Review of Public Administration* 6 (Winter 1983): 434-449.

Dorf, Ronald J., and Emerson, M. Jarvin. "Determinants of Manufacturing Plant Location for Nonmetropolitan Communities in the West North Central Region of the U.S." *Journal of Regional Science* 18 (No. 1 1978): 109-120.

Due, John F. "Studies of State-Local Tax Incentives on Location of Industry." *National Tax Journal* 14 (June 1961): 163-173.

Ecker, Deborah S., and Syron, Richard F. "Personal Taxes and Interstate Competition for High Technology Industries." *New England Economic Review* (September/October 1979): 25-31.

Erickson, Rodney A. "Business Climate Studies: A Critical Evaluation." *Economic Development Quarterly* 1 (February 1987): 62-71.

Fisher, James, and Hanink, Dean. "Business Climate: Behind the Geographic Shift of American Manufacturing." *Economic Review* 56 (6) June 1982.

Fox, William F. "Fiscal Differentials and Industrial Locations: Some Empirical Evidence." *Urban Studies* 18 (February 1981): 105-111.

———. "Local Taxes and Industrial Location." *Public Finance Quarterly* (1978): 93-114.

Gray, Ralph. "Industrial Development Subsidies and Efficiency in Resource Allocation." *National Tax Journal* 17 (2) (June 1964): 164-172.

Grieson, Ronald E. "Theoretical Analysis and Empirical Measurements of the Effects of the Philadelphia Income Tax," *Journal of Urban Economics* 8 (July 1980): 123-37.

Grieson, Ronald E. et al. "The Effect of Business Taxation on the Location of Industry." *Journal of Urban Economics*, April 1977, pp. 170-185.

Groves, Harold M., and Riew, John. "The Impact of Industry on Local Taxes — A Simple Model." *National Tax Journal* 16 (2) June 1963: 137-141.

Hale, Carl W. "The Optimality of Local Subsidies in Regional Development Programs." *Quarterly Review of Economics and Business* 9 (3) (Autumn 1969): 35-50.

Hartnett, Harry D. "Industrial Climate in Central Cities." *American Industrial Development Conference Journal* 7 (April 1972): 19-38.

Hekman, John S. "Survey of Location Decisions in the South." *Economic Review*, June 1982: 6-19.

Hellman, Daryl; Wassall, Gregory H.; and Escowitz, Herb. "The Role of Statewide Industrial Incentive Programs in the New England Economy." *New England Journal of Business and Economy* (1) 1973: 10-29.

Helms, L. Jay. "The Effect of State and Local Taxes on Economic Growth: A Time Series-Cross Section Approach." *The Review of Economics and Statistics*, 1985.

Hill, J, and Naroff, J.L. "The Effect of Location on the Performance of High Technology Firms." *Financial Management* 13 (Spring 1984): 27-36.

Hill, Lewis E. "Rates of Return on Municipal Subsidies to Industry: Comment." *Southern Economic Journal* 30 (April 1964): 358-359.

Kale, Steven R. "U.S. Industrial Development Incentives and Manufacturing Growth During the 1970's." *Growth and Change* 15 (January 1984): 26-34.

Laird, William E. and Rinehart, James. "Neglected Aspects of Industrial Subsidy." *Land Economics* 43 (1) (February 1967): 25-31.

Loewenstein, Louis K. "The Impact of New Industry on the Fiscal Revenues and Expenditures of Suburban Communities." *National Tax Journal* 16 (2) (June 1963): 113-136.

Luger, Michael I., and Shetty, Sudhir. "Determinants of Foreign Plant Start-Ups in the United States: Lessons for Policymakers in the Southeast." *Vanderbilt Journal of Transnational Law* 18 (Spring 1985): 223-245.

McDonald, J.F. "An Economic Analysis of Local Inducements for Business." *Journal of Urban Economics* 13 (1983): 322-336.

McGuire, T.J. "Are Local Property Taxes Important in the Intrametropolitan Location Decisions of Firms?" *Journal of Urban Economics* 18 (September 1985): 226-234.

———. "Interstate Tax Differentials, Tax Competition, and Tax Policy." *National Tax Journal* 39 (September 1986): 367-373.

Merenda, M.J. "An Empirical Investigation of Facility Location Decisions for New Hampshire: Executive Experiences and Perceptions." *New England Journal of Business and Economics* 8 (1982): 53-74.

Moore, F.T., and Peterson, J.W. "Regional Analysis: An Inter-Industry Study of Utah." *Review of Economics and Statistics* 37 (4) (November 1955): 368-383.

Morgan, William E., and Brown, Lee. "The Impact of State and Local Taxation on Industrial Location: A Measure for the Great Lakes Region." *Quarterly Review of Economics and Business*, Spring 1974, pp. 67-77.

Morgan, William E. and Hackbart, Merlin M. "An Analysis of State and Local Tax Exemption Programs." *Southern Economic Journal* 41 (2) (1976): 200-205.

Morse, G.W., and Farmer, M.C. "Location and Investment Effects of a Tax Abatement Program." *National Tax Journal* 39 (June 1986): 229-236.

Mulkey, David and Dillman, B.L. "Location and the Effects of State and Local Development Subsidies." *Growth and Change* 7 (April 1976): 71-80.

———. "Location and the Effects of State and Local Industrial Development Subsidies: A Reply." *Growth and Change* 8 (October 1977).

Newman, Robert. "Industry Migration and Growth in the South." *Review of Economics and Statistics* 65 (February 1983): 76-86.

Oster, Sharon. "Industrial Search for New Locations: An Empirical Analysis." *Review of Economics and Statistics*. May 1979, pp. 288-92.

Papke, James A., and Papke, Leslie E. "Measuring Differential State-Local Tax Liabilities and Their Implications for Business Investment Location." *National Tax Journal* 39 (September 1986): 357-366.

Plaut, Thomas, and Pluta, Joseph E. "Business Climate, Taxes and Expenditures, and State Industrial Growth in the United States." *Southern Economic Journal*, 50 (June 1983): 99-119.

Pluta, Joseph E. "Taxes and Industrial Location." *Texas Business Review* January/February 1980, pp. 1-6.

Poole, R.W. "An Approach for Evaluating the Impact of State-Local Taxes on Industrial Location." *New Mexico Business* 23 (June 1970): 5-11.

Rasmussen, David W.; Bendick, Marc, Jr.; and Ledebur, Larry C. "A Methodology for Selecting Economic Development Incentives." *Growth and Change* 15 (January 1984): 18-25.

Rasmussen, David W. et al. "Evaluating State Economic Development Incentives from a Firm's Perspective." *Business Economics* 17 (May 1982): 23-29.

Revzan, Lawrence H. "State and Local Tax Policies and Industrial Location Decisions." *Popular Government*, Winter 1976.

Revzan, Lawrence H., and Gutchess, Susan. "Assessing Business Incentives: Computer Analysis Can Forecast Effects." *Economic Development Commentary*, Spring 1984, pp. 14-18.

Rinehart, James R. "Rates of Return on Municipal Subsidies to Industry." *Southern Economic Journal* 29 (April 1963): 297-306.

———. "Rates of Return on Municipal Subsidies to Industry: Reply." *Southern Economic Journal* 30 (April 1964): 359-361.

Rinehart, James R. and Laird, William E. "Community Inducements to Industry and the Zero-Sum Game." *Scottish Journal of Political Economy* 19 (1) (February 1972): 73-90.

———. "Location Effects of State and Local Industrial Development Subsidies: Comment." *Growth and Change* 8 (4) (October 1977): 44-46.

———. "A Refinement of Local Industrial Subsidy Techniques." *Mississippi Valley Journal of Business and Economics* 3 (2) (Spring 1968): 90-99.

Romans, Thomas, and Ganti, Subrahmanyan. "State and Local Taxes, Transfers and Regional Economic Growth." *Southern Economic Journal* 46 (October 1979): 435-44.

Rubin, Barry M., and Zorn, C. Kurt. "Sensible State and Local Economic Development." *Public Administration Review*, March/April 1985, pp. 333-339.

Sahlings, Leonard. "Are State and City Corporate Income Taxes Stifling Investment in New York City?" *Federal Reserve Bank of New York*, December 1978.

Samaza, Gerald W. "A Benefit Cost Analysis of a Regional Development Incentive: State Loans." *Journal of Regional Science* 10 (1970): 375-396.

Spiegelman, Robert G. "A Method for Determining the Location Characteristics of Footloose Industries: A Case Study of the Precision Instrument Industry." *Land Economics* Vol. XL (1) (February 1964).

Steinnes, Donald N. "Business Climate, Tax Incentives and Regional Economic Development." *Growth and Change* 15 (1984): 38-47.

Stober, William J. and Falk, Lawrence H. "The Effect of Financial Inducements on the Location of Firms." *Southern Economic Journal* 36 (1) (July 1969): 25-35.

———. "Poorly Conceived Financial Inducements: A Study of Louisiana's Gas Severance Tax Rebates." *Social Science Quarterly* 51 (June 1970): 108-119.

Struyk, Raymond J. "An Analysis of Tax Structure, Public Service Levels and Regional Economic Growth." *Journal of Regional Science*, Winter 1967.

———. "Spatial Concentration of Manufacturing Employment in Metropolitan Areas." *Economic Geography* 48 (1972): 189-192.

Stutzer, Michael J. "The Statewide Economic Impact of Small Issue Industrial Revenue Bonds." *Quarterly Review* 9 (1985): 2-14.

"Survey of Current State Economic Development Programs," *Columbia Journal of Law and Social Problems* 17 (Summer/Fall 1983): 353-464.

Thompson, James H. "Local Subsidies for Industry: Comment." *Southern Economic Journal* 29 (October 1962): 114-119.

Vasquez, Thomas and deSeve, Charles. "State/Local Taxes and Jurisdictional Shifts in Corporate Business Activity: The Complications of Measurement." *National Tax Journal* 30 (3) (1977): 285-298.

Vedder, Richard K. "Rich States, Poor States: How High Taxes Inhibit Growth." *Journal of Contemporary Studies*, Fall 1982, pp. 19-32.

Warner, Paul D. "Business Climate, Taxes, and Economic Development." *Economic Development Quarterly* 1 (November 1987): 383-390.

Wasylenko, Michael. "Evidence of Fiscal Differentials and Intrametropolitan Firm Location," *Land Economics* 56 (August 1980): 339-49.

Wasylenko, Michael and McGuire, Theresa J. "Jobs and Taxes: The Effect of Business Climate on States' Employment Growth Rates." *National Tax Journal* 38 (December 1985): 497-511.

Watkins, Alfred. "Good Business Climates: The Second War Between the States." *Dissent* 27 (Fall 1980): 476-485.

Wheaton, W.C. "Interstate Differences in the Level of Business Taxation." *National Tax Journal* 36 (March 1983): 83-94.

Williams, William V. et al. "A Measure of the Impact of State and Local Taxes on Industry Location." *Journal of Regional Science* 7 (Summer 1967): 49-59.

Williams, William V. et al. "The Stability, Growth, and Stabilizing Influence of State Taxes." *National Tax Journal* 26 (2) (1973): 267-273.

3. Documents, Reports, Unpublished Papers, Theses, Etc.

Alexander Grant and Company. *A Study of Business Climate of the States*, Paper prepared for conference of State Manufacturers' Association, Chicago, Ill., 1979.

Ambrosius, Margery M. "Effects of State Development Policies on the Health of State Economies: A Time Series Regression Analysis." Paper delivered at the annual meeting of the Midwest Political Science Association, Chicago, Ill., April 10-12, 1986.

Apilado, V.P. "Public Administration of Financial Incentives in Industrial Plant Location: Industrial Aid Bonds." Papers in Public Administration 26. Arizona State University, Institute of Public Administration, 1973.

Aslan Associates. "Industrial Site Location Surveys." Report prepared for the Midwestern Legislative Conference Business Development Task Force, November 5-6, 1988.

Arthur D. Little, Inc. *An Evaluation of Nebraska's Industrial Development Marketing Efforts*. Report to the Nebraska Department of Economic Development. Cambridge, Mass.: Arthur D. Little, 1982.

———. *Massachusetts' Economy: The State's Competitive Position*. Cambridge, Mass.: Arthur D. Little, Inc., 1981.

Barnes, T.J., and Fitzsimmons, J.D. *Minnesota Manufacturing: A Comparative Geographic Study of Production*. Report to the Minnesota Business Partnership and the Center for Urban and Regional Affairs. Minneapolis: University of Minnesota, Department of Geography and School of Management, 1982.

Brooks, Stephen; Tannenwald, Robert; Sale, Hillary; and Puri, Sandeep. *The Competitiveness of the Massachusetts Tax System*. Boston: Special Commission on Tax Reform, 1986.

Burton, J.F. *Interstate Variations in Employers' Costs of Workmen's Compensation: Effect on Plant Location Exemplified in Michigan*. Kalamazoo, Mich.: W.E. Upjohn Institute for Employment Research, 1966.

Carbert, L.E. *The Impact of State and Local Taxes in North Carolina*. Raleigh: North Carolina Tax Study Commission, 1956.

Carlton, Dennis. "Birth of Single Establishment Firms and Regional Variations in Economic Costs." Report 7729. Chicago: Center for Mathematical Studies in Business and Economics, University of Chicago, 1977.

———. "Models of New Business Location." Report 7756. Chicago: Center for Mathematical Studies in Business and Economics, University of Chicago, 1977.

———. "Models of Single Establishment Births." SIC 3079, Report 7730. Chicago: Center for Mathematical Studies in Business and Economics, University of Chicago, 1977.

Conway, McKinley. "The Fifty Legislative Climates." In *Industrial Development and Site Location*. Atlanta: Conway Data, Inc., 1966-1985.

Cornia, G.; Testa, W.; and Stocker, F. "State-Local Fiscal Incentives and Economic Development." Urban and Regional Development Series No. 4. Columbus, Ohio: Academy for Contemporary Problems, 1978.

Digby, Michael F. "State Government and Economic Development: An Analysis and Evaluation of the Virginia Industrial Development Program." Ph.D. dissertation, University of Virginia, 1964.

Dorgan, B. "North Dakota's New Industry Tax Exemption: Economic Incentive or Tax Giveaway." A study for presentation to the State Board of Equalization and the North Dakota State Legislature. Fargo, N.D.: North Dakota State Tax Commission, 1980.

Economic Report to the Governor. Sacramento: State of California, 1977.

Erickson, R.A.; Miller, J.H.; and Wasylenko, M.J. *The Competitive Position of Pennsylvania Businesses.* Harrisburg: Business Council of Pennsylvania, 1983.

Fantus Co. *Economic Growth Assessment for the Western Marketing Region.* Report prepared for the Wisconsin Public Service Corporation, 1984.

———. *A Study of the Business Climate for the States,* Chicago: Fantus, 1975.

Fuller, Stephen S. and Towles, Joan E. "Impact of Intraurban Tax Differentials on Business and Residential Location in the Washington Metropolitan Area." Prepared for the D.C. Tax Revision Commission, 1978.

Genetski, Robert, and Chen, Young. "The Impact of State and Local Taxes on Economic Growth." Chicago: Harris Bank, 1978.

Grant Thornton. *The Annual Study of General Manufacturing Climates of the 48 Contiguous States of America.* Chicago: Grant Thornton International, 1979-1988.

Hansen, Susan B. "The Effects of State Industrial Policies on Economic Growth." Paper delivered at the 1984 annual meeting of the American Political Science Association, Washington, D.C., August 30 - September 2, 1984.

Kahn, J.P. "Report on the States," *Inc* 8 (1986): 57-66.

Ledebur, Larry. "Regional Location and Performance of Manufacturing in the United States." Washington, D.C.: The Urban Institute, October 1978 (unpublished).

Margolis, Nell, "Fifth Annual Report on the States," *Inc.* (October 1985): 90-98.

Minnesota Tax Study Commission. *1978 Comparative Business Climate Study.* Minneapolis: Minnesota Tax Study Commission, 1978.

Morgan, W. "The Effects of State and Local Tax and Financial Inducements on Industrial Location." Ph.D. dissertation, University of Colorado, 1964.

New York State Department of Commerce. *The Use of Public Funds or Credit in Industrial Location.* Albany: N.Y. State Department of Commerce Research Bulletin No. 6, October 1963.

Otto, Daniel and Stone, Kenneth. *Analysis of Economic Development Incentives for Iowa.* Iowa City: University of Iowa, Institute of Urban and Regional Research, 1982.

Padda, K. "Annual Report Card on the States." *Inc.* October 1981, pp. 90-98.

Papke, James A. "Investment-Tax Incentives as State Industrial Policy: Explorations Through Microanalytic Similarities." West Lafayette, Ind.: Purdue University, Center for Tax Policy Studies, Paper #3, April 1984.

———. "The Taxation of the Saturn Corporation: Intersite Microanalytic Simulations. West Lafayette, Indiana: Purdue University, Center for Tax Policy Studies, Paper #4, March 1985.

Papke, Leslie E. "The Measurement and Effect of Interstate Business Tax Differentials on the Location of Capital Investment." West Lafayette, Indiana: Purdue University, Center for Tax Policy Studies, Paper #5, October 1985.

Pennsylvania Economy League, *The Relative Tax Cost to Manufacturing Industry: A New Comparison of Pennsylvania With Several Other States.* Pittsburgh: Pennsylvania Economy League, 1957.

Posner, B.G. "A Report on the States: Inc.'s Second Annual Study Rates 50 Small Business Climates," *Inc.,* October 1982, pp. 95-100.

———. "*Inc.'s* Third Annual Report on the States." *Inc.,* October 1983, pp. 139-158.

Report of the Governor's Minnesota Tax Study Committee. St. Paul: State of Minnesota, 1956.

Rinehart, James. "Rates of Return on Municipal Subsidies to Industries." Ph.D. dissertation, University of Virginia, 1962.

Ross, W.D. "Tax Concessions and Their Effect." *Tax Institute of America Proceedings of 1957.* National Tax Association: 216-24.

Runyon, Kersteen, Ouellette and Co, *Maine Business Climate Satisfaction Survey.* Portland, Me.: Runyon, Kersteen, Ouellette and Co., 1986.

"Sixth Annual Report on the States." *Inc.,* October 1986, pp. 90-104.

Small, Kenneth A. "Geographically Differentiated Taxes and the Location of Firms." Princeton Urban and Regional Center Manuscript, 1982.

Sosnick, Stephen H. "The Local Tax Impact of a New Plant." Occasional Paper No. 1, Institute of Governmental Affairs, University of California-Davis, January 1964.

Stinson, T.F. *The Effects of Taxes and Public Finance Programs on Local Industrial Development.* Agricultural Economic Report No. 33, Economic Research Service, Washington, D.C.: U.S. Department of Agriculture, 1968.

Stockfish, J.A. *A Study of California's Tax Treatment of Manufacturing Industry.* Sacramento: California's Economic Development Agency, 1961.

Strategy for the Eighties: High Technology Industrial Development. Topeka: Kansas Department of Economic Development, 1982.

Sullivan, D. and Newman, R. "Econometric Analysis of Business Tax Impacts on Industrial Location: What Do We Know and How Do We Know It?" Working Paper, Miami University (Ohio), Department of Economics, 1983.

Wayslenko, Michael. "Business Climate, Industry and Employment Growth: A Review of Recent Evidence." Syracuse, N.Y.: Maxwell School of Citizenship and Public Affairs, Syracuse University, 1985.

———. "The Effects of Business Climate on Employment Growth in the States Between 1973 and 1980," Report for the Minnesota Tax Study Commission, 1984.

Wisconsin Department of Resource Development. *An Analysis of State and Local Industrial Development Incentives in Wisconsin, its Neighbor States and the Nation.* Madison: 1965.

Wolkenstein, Harry W. "The Unfavorable Consequences of Tax Concessions to Business Location and Development." *Proceedings of the 54th Annual Conference on Taxation.* National Tax Association, Harrisburg, Pa., 1961.

Wonnacott, R.J. *Manufacturing Costs and the Comparative Advantage of United States Regions,* Minneapolis: University of Minnesota Upper Midwest Economic Study, 1963.

Yankelovich, Skelly, and White, Inc. "A Survey of Business Executives' Attitudes Toward Wisconsin as a Business Location." Report prepared for the Wisconsin Department of Development, April 1984.

Yortema, D.B. *Michigan's Taxes on Business, 1956.* Lansing: Michigan Senate Tax Study Committee, 1959.

4. *Newspaper Articles, Magazines and Miscellaneous Publications*

Bergin, Thomas P. "Competition for Industry. Part 2: The South and the Use of Public Funds." *Municipal South,* December 1960.

Bergin, Thomas P., and Eagan, William F. "Are Subsidies Worthwhile?" *Industrial Development and Manufacturers Record* 129 (8) (July 1980): 77-78.

Biermann, W.W. "The Validity of Business Climate Rankings: A Test." *Industrial Development,* March/April 1984, pp. 17-25.

"Business Climates: Best is Southwest." *Electronic World,* July 1986, p. 56

"Business Climate as a Factor in the Industrial Location Decision." *Industrial Development,* January/February 1982, p. 37.

"Business Climates: Tallyho to the Dakotas." *Industry Week,* July 13, 1987, p. 24.

Carlson, Eugene. "Business-Climate Rating Stir Debate Over Study's Method." *The Wall Street Journal,* April 17, 1984, p. 33.

Center for Business and Economic Research. "The Estimated Economic Impact of Toyota on the State's Economy." Lexington, Ky.: University of Kentucky, 1986 (mimeographed).

Cochran, Wendell. "How Good is Iowa's Business Climate?" *Des Moines Register,* January 10, 1982, pp. 1Y and 3Y.

Couretas, J. "Climate Rankings Give Some States Chills." *Business Marketing,* June 1984, p. 15.

Duvall, Richard A. "Industry, States Rate Incentive, Assistance Programs." *Industrial Development,* November/December 1968, pp. 26-30.

Hagstrom, J. and Guskind, R. "Playing the State Ranking Game: A New National Pastime Catches On." *National Journal,* June 30, 1984, pp. 1268-74.

Harding, C.F. "Business Climate Studies: How Useful Are They?" *Industrial Development,* January/February 1983, pp. 22-23.

Hodge, J. "A Study of Industries' Regional Investment Decision." Unpublished paper, Federal Reserve Bank of New York, 1978.

Hotard, Kenneth N. "Tax Incentives for Business: The Myths and Realities." *State Legislatures,* September 1982, pp. 14-15.

"How Manufacturers Rank the States." *Chemical Week,* April 25, 1984, p. 20.

"Incentives: A Second Look - Survey of the Industrial Facility Planner's View of Special Incentives: An Update." *Industrial Development,* March/April 1984, p. 40.

"Iowans Say Change in Business Climate Needed." *IDC Digest* 10 (7) (July 1981).

Koretz, G. "Where to Find the Sunniest Business Climate." *Business Week,* April 13, 1987, p. 24.

McIntyre, Michael J. "Tax Incentives for Investment: A Review of a Study of the Studies." *Tax Notes,* March 9, 1981, pp. 491-492.

"Manufacturers Rank Sun Belt First for Plant Locations, But North is Closing Gap." *Journal of Accounting* 160 (August 1985): 44.

Moskal, B.S. "Business Climate: Florida Sunniest: States Near Great Lakes Rank Low." *Industry Week,* April 30, 1984, p. 18.

"Move to Delaware." *Fortune,* December 22, 1986, pp. 8-9.

National Governors' Association. "Survey of State Linkages Between Employment, Training, and Economic Development." In *Labor Notes,* July 1985a, pp. 15-17.

Norris, D.A., and Norris, J.M. "Places Rated Berated." *American Demographics,* March 1986.

Pilcher, Dan. "Assessing State Business Climates." *State Legislatures,* August/September 1983, pp. 9-12.

Purcell, H.I. "State and Local Taxes: A Significant Site Selection Variable." *Industrial Development,* November/December 1968, pp. 31-32.

"Quality of Life, Business Climate Factors." *Site Selection Handbook,* April 1985, p. 326.

Revzan, Lawrence H. "Enterprise Zones: Will They Affect Industrial Location Decisions?" *Industrial Development* September/October 1981, pp. 24-28.

Robinson, Roland I. "Subsidizing Industry at Everyone's Expense: Tax-Exempt Industrial Bonds." *Challenge: The Magazine of Economic Affairs* October 1963.

Souder, William. "Minnesota's Business Climate: Storm Warning Under Blue Skies." *Corporate Report: Minnesota* 12 (2) December 1981.

"South and West Gain From Tax Incentives and Growth." *Engineering News-Record*, February 4, 1982, p. 14.

"Study Rates States' Business Climates." *Automotive News*, February 22, 1982, p. 76.

Updegrave, W.L. "Ten Boomtowns You Can Bet On Year 2000." *Money*, November 1985, pp. 74-78.

"Warmer Climates: Southern Tier Rated Best for Business." *Industry Week* 226 (July 8, 1985): 21.

"Why Corporate America Moves Where." *Fortune*, Market Research Survey, New York: Times Inc., 1982.